CRADLE OF CHEMISTRY

CRADLE OF CHEMISTRY

*The Early Years of Chemistry
at the University of Edinburgh*

Edited by
Robert G. W. Anderson

First published in 2015 by
John Donald, an imprint of Birlinn Ltd

West Newington House
10 Newington Road
Edinburgh
EH9 1QS
www.birlinn.co.uk

ISBN: 978 1 906566 86 9

The publishers gratefully acknowledge the support of
The University of Edinburgh and Drs Alfred and Isabel Bader
towards the publication of this book.

British Library Cataloguing-in-Publication Data
A catalogue record for this book is available on request
from the British Library

Typeset in Garamond by
Koinonia, Manchester
Printed and bound in Britain by
TJ International, Padstow, Cornwall

Contents

List of Illustrations

Black and white plate section

1. Herman Boerhaave (1668–1738), professor of chemistry at the University of Leiden. By George White, published by Thomas Bowles senior, c. 1700–25. Mezzotint from the Fisher Collection, courtesy Chemical Heritage Foundation, Philadelphia. Photograph by Gregory Tobias.
2. Joseph Black's three main visualisations of affinity: (a) chiasm, (b) circlets, (c) a square affinity table. From Paul Panton's notes taken at Joseph Black's lectures, 1778, Chemical Heritage Foundation MS QD14.B533 1778 (chiasm: vol. 3, f. 107; circlet: vol. 6, f. 67; table: vol. 6, f. 17).
3. Experimental apparatus depicted alongside a set of instructions in Joseph Black's transcribed 1782 lecture notes. Royal Society of London.
4. (a) Charles Blagden's unsuccessful attempt to inscribe Black's affinity circlets into his student notebook. (b) Blagden's reconfiguration of Black's circlets into squares. Wellcome Library, London.
5. Venel, Gabriel François (1723–1775), 'Table des Rapports', *Cours de Chymie*. Wellcome Library, London.
6. Affinity table in Lewis, William, *The New Dispensatory* (London: Nourse, 1753), p. 11.
7. Leith-pattern wine bottle, from near Forres, Moray, late eighteenth century. Photograph copyright National Museums Scotland.
8. Nooth's apparatus for aerated mineral waters, deemed to be good for the health, c.1775. Photograph copyright National Museums Scotland.
9. Wedgwood's pyrometer, a temperature-measuring instrument for kilns, first described in 1782. Photograph copyright National Museums Scotland.

Colour plate section

Preface

It was on 9 December 1713 that Edinburgh Town Council, who controlled the University of Edinburgh, appointed James Crawford to a chair of physic (that is, medicine) and chymistry. A year later, on 15 December 1714, an advertisement appeared in the *Scots Courant* newspaper announcing that 'Dr. James Crawford, Professor of Medicine in the University of Edinburgh, begins a Course of Chimistry, within the College on the 16th Instant, at 2 in the Afternoon'. At that point the study of chemistry was formally born in Edinburgh, and the subject has flourished, without interruption, to the present day.

Anniversaries such as these provide the opportunity for reassessment. In June 2012, Eleanor Campbell, the current Professor of Chemistry, was considering with her colleagues how the 300th anniversary of Crawford's appointment might be celebrated. Following discussions with others, she decided that one element might be a symposium on the history of chemistry at Edinburgh. She assembled a small group consisting of herself, Professor Emeritus Robert Donovan, and Robert Anderson, Chairman of the Society for the History of Alchemy and Chemistry, to look into the feasibility of the idea. It appeared that there was a sufficient number of specialist historians to make a day's meeting a viable possibility. The Royal Society of Edinburgh was consulted and it was suggested that the meeting might form part of its annual programme. It was agreed that it should take place on 24 October 2013, a day close to the actual tercentenary. On consulting fellow historians, Robert Anderson found that there was enthusiasm for the idea, and invitations were issued.

The first century of chemistry at the University of Edinburgh was particularly important for development of the subject in Great Britain. A chair had existed from 1702 in Cambridge and there was sporadic teaching in Oxford in the eighteenth century, but in neither university was there the same level of excellence *and* continuity as at Edinburgh over three centuries. Glasgow's lectureship had been established in 1747, but the chair of chemistry did not arise until the next century. It was decided to concentrate the day's papers on the period of Edinburgh's pre-eminence, up to the middle of the nineteenth century. At this time, chemistry was inextricably bound up with the teaching of medicine, the Edinburgh faculty being set up in 1726. It has been estimated that during the second half of the eighteenth century, the Scottish universities (largely Edinburgh, but also Glasgow) taught 85% of all British medical graduates. The two English universities produced relatively few medical doctors, training generally being carried out in the London teaching hospitals. It was not until the University of London (later renamed University College, after other colleges were established) was founded in 1826 that chemistry became an important subject for study in the south, and from then for the next century it was nearly always Scots who occupied the chair of chemistry at University College London.

The symposium was by no means the sole element of the celebratory year. Events started at the end of January 2013 with a Burns Supper and Céilidh (it is noteworthy that Professor Joseph Black was a subscriber to Robert Burns's 1787 edition of *Poems, Chiefly in the Scottish Dialect*). In March, the student ChemSoc's Walker Memorial Lecture was delivered by Robert Anderson; in May the Tercentenary Debate, 'Science and Education in an Independent Scotland' was held; and Professor Sir Fraser Stoddart gave a public lecture in June on 'Chemistry's Place in Today's World'. Between August and November, the University Library hosted an exhibition, 'Edinburgh 300: Cradle of Chemistry', which was imaginatively curated by Andrew Alexander of the School of Chemistry. Perhaps the most innovative event was the writing and first performance, on 24 October, of a chemistry-themed opera, 'Breathe Freely', by the composer Julian Wagstaff. It concerned a Polish independence fighter and chemist, Stanislaw Hempel, who came to Edinburgh in 1943, and his relations with Professor James Kendall and Chrissie Miller, the first woman chemist to be elected a fellow of the Royal Society of Edinburgh. The opera was performed to a large and appreciative audience in the Assembly Rooms on the evening of the symposium.

The symposium was opened with a short and entertaining address by the President of the Royal Society of Edinburgh, Sir John Arbuthnott, to whom we are most grateful for the provision of the Society's rooms (two of the chemistry professors celebrated at the meeting had been founder-fellows of the Society in 1783). Also to be thanked for their participation are the symposium's two chairs, Professor Lesley Yellowlees (Professor of Inorganic Electrochemistry, Vice-Principal and Head of College of Science and Engineering, University of Edinburgh, and first woman President of the Royal Society of Chemistry) and Professor Ewen Cameron, Sir William Fraser Professor of Scottish History and Palaeography at the University of Edinburgh.

It gives pleasure to thank those who sponsored the symposium and the production of this volume of all the papers which were presented, without whom it would not have happened: Drs Alfred and Isabel Bader, The Alembic Club, The Society for the History of Alchemy and Chemistry, Bruker, UCB, Infineum, and the University of Edinburgh, including the School of Chemistry itself. Staff of the Royal Society of Edinburgh were most helpful in the organisation of the symposium. Thanks are also due to the publisher, Birlinn, for the advice and help they have provided. Finally, thanks must be offered to all the contributors, who have brought awareness of the significance of the early days of the Edinburgh chemistry school to us all.

Eleanor Campbell
Robert Donovan
Robert G. W. Anderson

Notes on Contributors

Tom Addyman

tomaddyman@addyman-archaeology.co.uk
Tom Addyman studied Archaeology and Anthropology at the University of Cambridge before training at the Institute of Advanced Architectural Studies at the University of York. In 1997 he established Addyman Archaeology, now based at Leith, one of the leading archaeological companies in the north of Britain concerned with historic building, recording and analysis, and since 2006 a division of Simpson and Brown Architects of which he is a partner. He has extensive experience as an archaeologist and researcher of historic buildings in the UK and internationally, including work in Central America on archaeological sites of the Maya, on colonial period and Native American sites in the United States and Puerto Rico, and on colonial and Sikh period sites in India.

Andrew J. Alexander

andrew.alexander@ed.ac.uk
Dr Andy Alexander is a Senior Lecturer in Chemical Physics at the University of Edinburgh. Educated at Edinburgh, Oxford, and Stanford Universities, he returned to Edinburgh to take up a Royal Society University Research Fellowship and Lectureship. His research interests are photochemistry and nucleation. Dr Alexander co-ordinated events for the tercentenary celebrations in 2013 and curated an exhibition titled 'Cradle of Chemistry' at the University Main Library.

Robert G.W. Anderson

rgwa2@cam.ac.uk
Robert Anderson, an Oxford graduate, spent his career working in national museums, latterly directing the National Museums of Scotland and then the British Museum. His first post was at the Royal Scottish Museum, Edinburgh, where he was responsible for the collection of

historical chemical apparatus and instruments. In 1978 he published *The Playfair Collection*, a catalogue and commentary on the material transferred in 1858 to the Museum from the University chemistry laboratory. He has published mainly on the history of chemistry, and also on museums and their collections. He co-edited *The Correspondence of Joseph Black*, which appeared in 2012. In 1986 he received the ACS Dexter Award. He is Chair of the Society for the History of Alchemy and Chemistry, Vice-Chair of the Chemical Heritage Foundation, Philadelphia, and an Emeritus Fellow of Clare Hall, Cambridge.

Hasok Chang

hc372@cam.ac.uk

Hasok Chang is the Hans Rausing Professor of History and Philosophy of Science at the University of Cambridge. He received his degrees from Caltech and Stanford, and has taught at University College London. He is the author of *Is Water H$_2$O? Evidence, Realism and Pluralism* (2012), and *Inventing Temperature: Measurement and Scientific Progress* (2004). He served as the President of the British Society for the History of Science from 2012 to 2014.

John R.R. Christie

jrrc_@hotmail.com

John R.R. Christie read History at the University of Edinburgh, graduating in 1969. After postgraduate research at Edinburgh he became first a Research Fellow, then Lecturer in History and Philosophy of Science at the University of Leeds, where he continued his career until 2007, when he moved to Oxford, becoming a member of the Faculty of History. He has held Research Fellowships at Edinburgh's Institute for Advanced Studies and at the Max Planck Institute for History of Science, Berlin, to which he returns in 2015. In 2017 he will be the Cain Conference Fellow at the Chemical Heritage Foundation, Philadelphia. His research has focused on eighteenth-century chemistry in Scotland, and latterly, in England, with particular attention to the work of Joseph Priestley.

Matthew Daniel Eddy

m.d.eddy@durham.ac.uk

Matthew Daniel Eddy is Senior Lecturer in the History and Philosophy of Science at Durham University, England. He is a historian of modern Europe, with particular interests in the intellectual, social and religious history of Britain and its colonies. He is author of *The Language of Mineralogy: John Walker, Chemistry and the Edinburgh Medical School, 1750–1800* (2008), co-editor of *Chemical Knowledge in the Early Modern World* (2014) and is currently writing a book that offers a new visual approach to the intellectual history of the late Enlightenment.

John Henry

John.Henry@ed.ac.uk

Professor Emeritus John Henry has just retired from the University of Edinburgh, where he was Director of the Science Studies Unit. He has published widely in the history of science, from the sixteenth century to the nineteenth, and from atomism to palaeontology. He is also the author of a successful textbook, *The Scientific Revolution and the Origins of Modern Science*, 3rd edition (Palgrave Macmillan, 2008).

Peter J.T. Morris

peter.morris@sciencemuseum.ac.uk

Dr Peter J.T. Morris is Keeper of Research Projects in the Research & Public History Department at the Science Museum. He applied to read chemistry at Edinburgh but ended up at Oxford, where he wrote an undergraduate thesis on eighteenth-century chemical education. Morris has published books on the history of synthetic rubber and polymers, modern chemical instrumentation, and the work of Robert Burns Woodward. He was given the Edelstein Award for the history of chemistry by ACS HIST Division in 2006. His latest book – a history of the chemical laboratory from the 1590s to 2004 – has been published by Reaktion Books.

A.D. Morrison-Low

A.Morrison-Low@nms.ac.uk

Alison Morrison-Low won the 2008 Paul Bunge Prize for her book *Making Scientific Instruments in the Industrial Revolution* (Aldershot: Ashgate, 2007). She is Principal Curator, Science, at National Museums Scotland.

John C. Powers

jcpowers@vcu.edu

John C. Powers is an Associate Professor in the Department of History and Assistant Director of the Science, Technology, and Society Program at Virginia Commonwealth University. He published *Inventing*

Chemistry: Herman Boerhaave and the Reform of the Chemical Arts (University of Chicago Press, 2012), which examined the pedagogical and philosophical roots of Boerhaave's chemical work within the University of Leiden's medical faculty. His latest research focuses on the introduction of thermometry into chemistry and its impact on pedagogy, theoretical development and artisanal practices.

Georgette Taylor

georgette.taylor@gmail.com

Georgette Taylor's 2006 PhD on the chemical affinity theories of the eighteenth century explored the teaching of chemistry, in particular by William Cullen and Joseph Black and by many of their ex-students. She won the 2008 Partington Prize awarded by the Society for the History of Alchemy and Chemistry for 'Tracing Influence in Small Steps: Richard Kirwan's Quantified Affinity Theory'. A post-doctoral fellowship followed, with the project 'Analysis and Synthesis in 19th-Century Chemistry: Towards a New Philosophical History of Scientific Practice'.

ONE

Introduction

ROBERT G.W. ANDERSON

The teaching of chemistry at the University of Edinburgh came about through its association with medicine, and the ambition to establish a medical school which would enjoy a high reputation, would encourage Scottish students to stay at home rather than be attracted to foreign universities on the European Continent, while at the same time bring foreign students to Edinburgh.[1] It is clear that there were powerful aristocratic forces behind this plan, even if the details are not known.[2]

The first professor of medicine and chemistry, James Crawford, was appointed in 1713, some 13 years before the faculty was established. The second professor, Andrew Plummer, before being officially appointed, started to teach medicine in Edinburgh extramurally from 1724 with three colleagues, all of whom had attended the Leiden medical school. This was part of a thoroughly thought-out plan, and each part of the jigsaw fell into place two years later when the four of them were appointed founder-professors of the new medical faculty, together with the anatomist Alexander Monro *primus*. The cost to Edinburgh, through its City Council (which was the Patron to the University) was small: none of the faculty was salaried and the medical professors had to rely on student fees, private practice and, in the case of Plummer and his colleagues, entrepreneurship through the large-scale production of pharmaceuticals.[3] This had a long-term effect on the development of teaching, including chemistry, where establishing a high level of student registration was important to ensure personal income.

In addition to its more theoretical component, chemistry is a practical subject; experimentation and demonstration play an important role in its pedagogy. The popularity of their courses meant that the professors made frequent bids for improved accommodation and for supplies of apparatus. Some items had to be large so that students in the back of the crowded lecture hall could see what was being demonstrated. Though students could undertake a course in chemistry successfully without handling any apparatus at all (even though a dissertation was required of them before they could graduate), they observed a long series of experiments performed in front of them by their professors. Several professors developed considerable reputations for the brilliance of their demonstration technique, some viewers commenting that their chemical shows verged on the theatrical. Chemistry was becoming a fashionable pursuit all over the country, and a significant proportion of the Edinburgh audience – at some periods a majority of those registering – did not matriculate, and had no intention of becoming doctors.[4] The content of the lecture series after Andrew Plummer's time (he died in 1756) was not particularly geared to medicine. Plummer's successors, William Cullen and Joseph Black, promoted the more general application of chemistry,[5] especially geared to improving agriculture and nurturing industrial processes. Additional private courses for specialised groups and for the general public were given by them and by the next professor, Thomas Charles Hope. Despite the development of its theoretical content, relatively little was done to encourage research, and in the nineteenth century, the small cohort of Edinburgh students who wanted to take up chemistry as a profession began to look elsewhere for places to develop their practical technique after graduation.

The story of Edinburgh chemistry within the medical faculty is told in the chapters which follow. The formula which was developed worked well, and students from England, Ireland and the colonies, even from France, Germany, America and Russia, streamed into Edinburgh to study medicine. During the second half of the eighteenth century, Edinburgh had become the pre-eminent place to study chemistry. But other universities and nations would follow and catch up, even surpass. By quite early in the nineteenth century, France and several German states had diversified the concept of chemistry teaching and they too were attracting students from far and wide (including a few from Scotland).[6] But the change was not rapid. In Britain, Glasgow University showed distinct liveliness from time to time, and from the 1790s, the democratic (at least, cheaper) option of Glasgow's Andersonian Institution became available. Aberdeen and St Andrews never challenged the position of the lowland Scottish universities, though occasionally medical degrees were awarded, sometimes in dubious circumstances. Oxford and Cambridge's record was distinctly patchy; at this period, no formal education for undergraduates would lead to degrees in science and medicine in either place. The popular Scottish extramural schools, especially those established in Edinburgh, would often flourish under a clever teacher and must not be ignored when considering the overall picture of chemical education; they trained very large numbers.[7] Likewise, the London teaching hospitals need consideration, where many of the Oxbridge graduates would train to become doctors. At the end of the period covered by this survey, University College London (initially called London University), founded in 1826, would in many respects take the lead in science teaching.[8] UCL chemistry was not tied to a medical faculty and dealt with the subject in its own right. It is interesting to note that most of those appointed to the senior positions in chemistry in London over this period were either Scottish, or Scottish-trained, chemists.

This volume starts with John Henry providing a context for Scottish science at the beginning of the eighteenth century, concentrating mainly on the influence of Sir Isaac Newton. He suggests that Newton's ideas were not so much passively absorbed by the Scottish intellectual world, but that some Scots were fired by the ideas of Newtonianism, and it was they who were responsible for its wider diffusion. Indeed, Scotland occupied a central place in European culture in the eighteenth century (the term 'Scottish Enlightenment' has been a commonly used one) and it did not pass unrecognised; the poet and author Tobias Smollett would refer to Edinburgh as being 'a Hot-bed of Genius'. Leiden in the Netherlands had initially been a powerful influence on the development of Edinburgh medicine and chemistry. This story is taken up by John C. Powers in considering the influence of the Leiden professor Herman Boerhaave on James Crawford and Andrew Plummer, the first two professors of chemistry in Edinburgh. Both had been to Leiden – Plummer to prepare for a medical degree, Crawford earlier, studying for only a matter of weeks. But the figure of Boerhaave and his teaching would loom over Edinburgh for many years, and much of the earlier story of Edinburgh chemistry is concerned with how professors initially continued, and then broke away from his pedagogical model.

Georgette Taylor particularly considers the chemical pedagogy of Plummer and his successor, William Cullen. Plummer seems to have concentrated his teaching largely on drug preparation. Cullen did not admire this (in 1755, Cullen reassured the young Joseph Black, who was just about to start teaching, 'you need not be anxious, provided your course be better than Plummer's which it is impossible for it not to be'). However, Taylor does make it clear that there is a paucity of evidence about Plummer's course; a great deal more is known about Cullen's teaching. Relatively few science historians have viewed teaching from the point of view of the students, but this is what John Christie does in his chapter in considering the 1770s and 1780s in Edinburgh. It was in this period that the new chemistry of the French chemist Antoine-Laurent Lavoisier was being viewed cautiously by professors, but embraced enthusiastically by Edinburgh students. Unusually for the time, students themselves formed their own scientific societies, the Chemical Society being the earliest such society anywhere, and it was at its meetings that Lavoisier's ideas were being debated. Indeed, Black wrote to Lavoisier to tell him that Edinburgh was a place 'where the students enjoy the most perfect liberty in chuseing their philosophical opinions'.

Matthew Eddy is also concerned about the detail of teaching method. He considers how students in Enlightenment Scotland learned the forms, skills and routines that would later, as diagrams on paper, transform the description of the natural world. Black's diagrams, found reproduced in many sets of student lecture notes, were used to convey ideas about chemical attraction between substances. The ideology adopted was that some substances had stronger attraction for certain substances than for others; these 'affinities' were classed as single or double, according to the competing attractions in mixtures of the substances. Black's diagrams were innovative, intended to be accessible and useful to students. They are frequently to be found in manuscript notebooks personally owned by the students.

Tom Addyman has looked at chemistry in Edinburgh from the point of view of its material culture. As an archaeologist, he dug near the site of Black's 1781 teaching laboratory in 2010 and 2011, and found a wide range of chemical artefacts and chemical substances. These had been buried in around 1820, when the Old Library of 1642 was demolished on top of the chemistry store. Presumably these materials were not considered to be worth saving. This contrasts with Alison Morrison-Low's paper: her interest lies mainly in the remarkable collection of 75 intact chemical wares from the Edinburgh teaching collection, which were transferred by Lyon Playfair in 1858 to the Industrial Museum of Scotland (now the National Museums of Scotland). She describes how Black probably obtained his glassware from Archibald Geddes, who ran one of the Leith glasshouses and who had attended Black's lectures.

The paper which follows, by Peter Morris, considers Black's dwelling house in Nicolson Street. After living in relatively modest quarters, in 1781 Black moved to a smart new house not far from the Old College. Morris shows that up to now, the wrong house has been identified as Black's, the right house having been unfortunately demolished in 1982 after a fire.

Robert Anderson follows these Black papers by considering the teaching of Black's successor, Thomas Charles Hope. He has been considered a dull teacher up to now (in spite of Charles Darwin's laudatory remarks about him), but on examining the notes from which he lectured, which survive in Edinburgh University Library, there is no denying his conscientiousness and efforts to remain up to date in his subject. Hope modelled himself on Black, in that he saw himself first and foremost as an educator, not as one whose job was to promote research activity in his students. His classes grew to a huge size: in 1823, some 559 registered for his annual course of chemistry.

The broadest sweep of the conference papers is offered by Andrew Alexander, who takes the story from William Gregory (who followed Hope as professor in 1844), via Lyon Playfair, to Alexander Crum Brown, who held the chair until 1908. But he points out that chemistry is not all about professors: in particular, Alexander concentrates on Archibald Scott Couper, a remarkable Edinburgh chemist who was an assistant to Playfair and who independently proposed the tetravalence of carbon. Sadly, Couper suffered from schizophrenia, and his working career was cut tragically short. A figure more widely known than any of the professors was a medical student who studied chemistry under Crum Brown in the 1876–7 session: Arthur Conan Doyle, creator of the Sherlock Holmes stories. The chemistry he learned during his medical studies is much in evidence in these, as Alexander points out.

Drawing proceedings to a close in his Afterword is Hasok Chang. Rhetorically, he asks two questions: 'What enabled the flourishing of chemistry in Edinburgh in the Enlightenment?' and 'Was there something distinctive about Edinburgh chemistry?' He turns his thoughts to the topic of the chemical research which was conducted in this period, wondering why little has been said about it. Finally, in considering these big issues, he asks why a working scientist today should pay attention to all this history. 'What is to be gained: curiosity and amusement, or a sense of heritage and context?' He concludes that its real value is that it opens minds.

This volume contains the closest investigation of a period of chemical history as seen through the medium of a teaching institution. If the papers contained in this volume encourage debate (and possibly disagreement), then the conference held in November 2013 will have been well worth the efforts made by its organisers.

Notes and References

1 Morrell, Jack B., 'The Edinburgh Town Council and its University 1719–1766' in Anderson, R.G.W. and A.D.C. Simpson (eds), *The Early Years of the Edinburgh Medical School* (Edinburgh: Royal Scottish Museum, 1976), pp. 1–26.

2 Emerson, Roger, *Academic Patronage in the Scottish Enlightenment: Glasgow, Edinburgh and St Andrews Universities* (Edinburgh: Edinburgh University Press, 2007).

3 Anderson, R.G.W., *The Playfair Collection and the Teaching of Chemistry at the University of Edinburgh 1713–1858* (Edinburgh: Royal Scottish Museum, 1976), pp. 5–10.

4 Rosner, Lisa, *Medical Education in the Age of Improvement: Edinburgh Students and Apprentices 1760–1826* (Edinburgh: Edinburgh University Press, 1991).

5 Donovan, Arthur, *Philosophical Chemistry in the Scottish Enlightenment* (Edinburgh: Edinburgh University Press, 1970).

6 Morrell, Jack B., 'The Chemist Breeders: The Research Schools of Liebig and Thomas Thomson', *Ambix* 19 (1972), pp. 1–46.

7 Anderson, Robert G.W., 'Chemistry Beyond the Academy: Diversity in Scotland in the Early Nineteenth Century', *Ambix* 57 (2010), pp. 84–103.

8 Davies, Alwyn and Peter Garrett, *UCL Chemistry Department 1828–1974* (St Albans: Science Reviews, 2013).

TWO

Science in the Athens of the North: The Development of Science in Enlightenment Edinburgh

JOHN HENRY

The Scottish Enlightenment was a truly remarkable historical phenomenon. While it is always possible for historians to show the continuities underlying historical change, there can be no denying that the many causative factors which combined together to give rise to the Scottish Enlightenment resulted in a marked sea change in Scottish culture – a notable disjunction from earlier history. Suddenly, Scotland found itself taking a central place in European life and letters. Even David Hume (1711–1776) was surprised:

> Is it not strange that at a time when we have lost our Princes, our Parliaments, our independent government, even the Presence of our chief Nobility, are unhappy in our accent and pronunciation, speak a very corrupt Dialect of the Tongue which we make use of, is it not strange, I say, that in these Circumstances, we shou'd really be the People most distinguished for Literature in Europe?[1]

Tobias Smollett's description of Edinburgh in the eighteenth century as a 'hot-bed of genius' is well known; William Smellie recounts the story of Mr Amyat, the king's chemist, standing in Parliament Square and breathlessly claiming: 'Here I stand at which is called the *Cross of Edinburgh*, and can, in a few minutes, take fifty men of genius and learning by the hand.'[2] These must have been remarkable times.

As has been hinted, the causes of this phenomenon are many and varied, and it is not necessary to go into them all here. But a focus on one of those causes is called for – the rise of the natural sciences as a widely recognised source of intellectual and cultural authority. Historians of science might be expected to extol the historical importance of science, but it is worth noting that even historians who do not normally pay any attention to developments in the sciences are forced to acknowledge the importance of science in the European-wide phenomenon of the Enlightenment. It is no coincidence that the Enlightenment followed hard on the heels of the period known to historians as the Scientific Revolution. Norman Hampson, in his classic 1968 study of the Enlightenment, pointed out that at the beginning of the seventeenth century, 'the cultural horizon of most educated men in Western Europe [. . .] was dominated by two almost unchallenged sources of authority: scripture and the classics'.[3] During the course of the seventeenth century, the natural sciences emerged as a potentially new and powerful source of authority – it was in the Enlightenment that that potential became fulfilled, and the sciences came to be recognised as the most reliable and certain means of discovering and establishing the truth.[4]

If a starting date for the Scientific Revolution were to be proposed, a good year to choose would be 1543 – this was the year that Nicolaus Copernicus published his *On the Revolutions of the Heavenly Spheres*, in which he insisted that the Earth was not stationary at the centre of the world system, but moved around the Sun along with all the other planets; and it was the same year that Andreas Vesalius transformed the study of anatomy by performing his own dissections and carefully recording what he actually saw, even if it meant deviating from the ancient authorities.[5] But more importantly, it can be noted that the end of the Scientific Revolution was effectively marked by the triumphant work of Isaac Newton. In just two land-mark books, Newton seemed to bring the major lessons of the Scientific Revolution together and to demonstrate the new power of the physical sciences. His *Principia mathematica* of 1687 was seen as an unassailable demonstration of how mathematics could be used to help an understanding of the workings of the physical world, while the *Opticks* of 1704 was seen as the perfect exemplar of the experimental method. Taken together, these books, as Jean d'Alembert wrote in the 'Preliminary Discourse' to that great monument of the Enlightenment, the French *Encylopédie* (1751), 'gave philosophy a form which apparently it is to keep' – and indeed so it proved, at least until the turn of the nineteenth into the twentieth century.[6]

Newton died in 1727, but his presence throughout the Enlightenment is palpable; so much so that he is often seen as the embodiment of the Age of Reason. Sir Isaiah Berlin, in a brief introduction to the Enlightenment written in 1956, even went so far as to say that if we wish to understand the European-wide Enlightenment we need to acknowledge 'Newton's influence as the strongest single factor'.[7] Sir Isaiah's authoritative pronouncement can be taken as a cue for focusing on Newtonianism as a major element in the Scottish Enlightenment, and in particular the Enlightenment in Edinburgh.

It might be assumed that Newton's fame and influence would spread first of all from Cambridge, or perhaps from London, and that it would only be felt in Edinburgh and other Scottish university towns after a significant delay. But this was very definitely not the case. It is a remarkable fact that Newtonian influence almost seems to emanate first of all from Edinburgh.[8] The chief figure in the beginning of this story is David Gregory (1659–1708), who succeeded his uncle, James Gregory (1638–75), as professor of mathematics at Edinburgh in 1683. His uncle had been a fellow of the Royal Society and had foreshadowed some of Newton's mathematical achievements on the way to the discovery of calculus (fluxions), and had also designed a reflecting telescope which Newton subsequently improved upon and built. Because of this family background, David Gregory was already aware of Newton's reputation as a mathematician even before he published the *Principia*, and when the *Principia* appeared, Gregory was one of the first to read it, and one of the few who could understand it. Although Gregory's teaching at Edinburgh was at too elementary a level to include Newtonian physics, he did introduce a number of his more gifted students to Newtonianism. He also introduced his good friend, Archibald Pitcairne (1652–1713), physician and nominal professor of medicine at the university, to Newtonian ideas.[9]

Gregory moved to Oxford to take up the Savilian Chair of Astronomy in 1691 – a position which he won with Newton's support. Interestingly, Newton favoured Gregory over his own friend Edmond Halley (1656–1742), who also applied for the Savilian Chair. It is usually said that Newton chose to support Gregory because he disapproved of Halley's godlessness, but Gregory's own lack of religion was well known – Halley had been described by a Scottish visitor to London as the only man 'that has less religion than Dr Gregory', which does not say much for Gregory's devotions.[10] It is perfectly possible, therefore, that Newton supported Gregory because he was the better mathematician – which he certainly was. At any rate, the first Newtonian among the Oxford professors came from Scotland.

So did the second: one of Gregory's brightest students in Edinburgh, John Keill (1671–1721), followed his master down to Oxford, and after developing ways of expounding Newtonian principles by experimental demonstrations in his rooms at Balliol College, was appointed as a lecturer in experimental philosophy at Hart Hall. He offered the first course on Newtonian natural philosophy, and the first reputedly based on 'experimental demonstrations', at either of the English universities. Furthermore, the published version of his lectures (*Introductio ad veram physicam*, Oxford, 1701) was the first systematic and simplified account of Newtonian physics. The innovatory nature of his teaching, based on experimental demonstration, was continued after 1710 by his student John Theophilus Desaguliers (1683–1744). Keill succeeded to the Savilian Chair after the death of Gregory (1708), and the death of Gregory's time-serving successor, in 1712.[11] Keill published a second set of his lectures, this time on astronomy, in 1718 (*Introductio ad veram astronomiam*). The populist intentions of this can be seen from the fact that when it appeared in English translation in 1721, Keill announced in the dedication that it was published 'at the Request, and for the Service of, the Fair Sex'. He went on to say that 'It was no Flattery to the Ladies, to say, that such of them as delight in Arts and Sciences, as to the Quickness of Perception and Delicacy of Taste, are equal, if not superior to Men.' Much has been made by historians of science in recent years of Francesco Algarotti's *Newtonianismo per le dame*, published in 1737, and seen as a pioneering work of Newtonian popularisation, but they have all overlooked Keill's *Introduction to the true astronomy*, published nearly two decades earlier.[12] There can be no denying, however, that Gregory and Keill were among the first to spread the Newtonian word, first in Edinburgh, and then more widely in Great Britain and on the Continent. Keill in particular, because of his publications, was instrumental in shaping the interpretation of Newtonianism among continental natural philosophers.[13]

Meanwhile, David Gregory's Edinburgh friend Archibald Pitcairne was developing and promoting a Newtonian version of medical theory and practice.[14] The mechanical philosophy of René Descartes (1597–1650) had been extended to cover human bodies and the way they break down, thereby providing a mechanical medical philosophy, or iatro-mechanistic philosophy. Similarly, followers of Galileo in Italy, Giovanni Borelli (1608–1679) and Lorenzo Bellini (1643–1704), had developed mechanistic conceptions of the body.[15] The young Pitcairne, like many of his contemporaries in medicine, saw iatromechanism as an exciting new theory that was clearly related to the latest, cutting-edge natural philosophy. Converted to Newtonianism by Gregory, with whom Pitcairne worked through the *Principia*, Pitcairne became the first Newtonian iatromechanist. But he was by no means the last. Perhaps because his medical philosophy seemed to be up to the minute, Pitcairne acquired a formidable reputation and was courted by the University of Leiden in the Netherlands, already the leading school of medicine in Europe. He was offered the chair of the practice of medicine in 1691, and journeyed to the Netherlands via Cambridge, where Newton gave him a copy of his important unpublished treatise *On the Nature of Acids* (*De natura acidorum*). Pitcairne needed no further encouragement, and in his inaugural lecture at Leiden he began his campaign to develop a theory and practice of medicine based on Newtonian principles. The authorities at Leiden were suitably impressed and immediately increased his salary, and he soon moved into the theory of medicine, teaching that course alternately with the official professor in that subject, the Cartesian iatromechanist, Charles Drélincourt (1633–1697). Pitcairne did not stay long in Leiden, resigning his chair at the end of 1693. But it is evident that he made his mark. While there he taught William Cockburn (1669–1739) and Richard Mead (1673–1754), who became leading Newtonian physicians, and the surgeons Robert Eliot (*fl.* 1705), who became first professor of anatomy at the University of Edinburgh in 1705, and John Monro (1670–1740), father of Alexander Monro *primus* (1697–1767), who built the Edinburgh Medical School into such a flourishing concern from 1726 onwards.[16]

Leiden was soon to become the most influential school of medicine in Europe thanks to the teaching of Herman Boerhaave (1668–1738). Although he was not taught by Pitcairne, there can be no denying that Boerhaave was very much a Newtonian in his theory of medicine. It is not clear where or when Boerhaave picked up his admiration for Newton, but it is perfectly possible that he learned of it from Drélincourt, or from others at Leiden who were still impressed by Pitcairne's teaching.[17] Boerhaave figures hugely in the success of Newtonianism in the Enlightenment, but although he is Dutch, he can still be claimed as being at least partially influenced by the Newtonianism that had come out of Edinburgh with Archibald Pitcairne.

Another major student of Pitcairne's was George Cheyne (1672–1743), who not only promoted Pitcairne's brand of Newtonian medicine but also tried his hand at expounding the theories of Newton for a wider audience – first with a simplified account of calculus (*Fluxionum methodus inversa*, 1703), and then with an early example of Newtonian natural theology – that is to say, trying to prove the existence of God by reference to the design seemingly inherent in the Newtonian system (*Philosophical Principles of Natural Religion*, 1705).[18]

It is worth pointing out that David Gregory, John Keill and George Cheyne had all produced accounts of Newtonian philosophy intended for a wide audience by 1705. The only English author to have attempted something similar was Samuel Clarke (1675–1729), but he did this in footnote annotations added to successive editions (1697, 1702, 1710) of Jacques Rohault's *System of Natural Philosophy*, a Cartesian textbook in use at the University of Cambridge. Clarke's footnotes were dictated by Rohault's text and did not provide a continuous and systematic account.[19] William Whiston was the first English writer to produce a systematic popular account of Newtonianism, but not until 1716.[20]

That the Newtonianism being promoted in Edinburgh at this time was eagerly taken up by students can be inferred from the complaint of the political theorist Alexander Fletcher of Saltoun (1653–1716), writing in 1704, that students were ignoring moral and civic instruction in favour of the sciences:

> by the present manner of education, the minds of young men are for many years debauched from all that duty and business to which they are born; and in the place of moral and civil knowledge and virtue, addict themselves to mathematical, natural, metaphysical speculations, from which many are never able to withdraw their thoughts.[21]

A little later, another highly gifted Scottish student of mathematics found his own way to mastery of Newtonianism. Colin Maclaurin (1698–1746) was educated at Glasgow University (taught by the Newtonian mathematician, Robert Simson)[22] and in 1713 defended a thesis, 'On gravity and other natural forces'. The following year he composed a remarkable moral treatise entitled 'On the good-seeking forces of the mind' (*De viribus mentium bonipetis*), which tried to analyse moral behaviour in mathematical, and Newtonian, terms.[23] Maclaurin became the youngest professor at any university when he became professor of mathematics at Marischal College, Aberdeen in 1717, at the age of 19. In 1726, however, thanks to the intervention of Newton himself, Maclaurin became professor of mathematics in Edinburgh and went on to consolidate his position as the leading Newtonian mathematician in Britain. Maclaurin already had an international reputation, but he cemented it in 1742 with his *Treatise of Fluxions*, which not only showed the uses of calculus but also defended it from criticisms that it was philosophically unsound, being based on inconceivable entities like infinitesimals.[24] Maclaurin is also the author of the best popular account of Newtonian physics, *An account of Sir Isaac Newton's philosophical discoveries*, which he finished by dictation from his deathbed, and which was published posthumously in 1748.[25]

It is fair to say that by the time of his death Maclaurin had long since established himself as the leading Newtonian in Enlightenment Britain. And he also figures largely in the history of the founding of the Royal Society of Edinburgh – or at least the successive precursor bodies from which it eventually emerged, the Medical Society of Edinburgh, and the Edinburgh Society for Improving Arts and Sciences and particularly Natural Knowledge.[26]

But Newtonianism was not just confined to the physical sciences. One of the most characteristic aspects of the Scottish Enlightenment is the development of the so-called Science of Man. The Science of Man itself soon branched out into different sub-disciplines – political economy, psychology, sociology, anthropology and so forth – but all of them were seen by their early proponents as extending the methods of Newtonian science, proven to be so successful in physics, to the human sciences. This is most famously noted in David Hume's self-professed claim that his *Treatise of Human Nature* of 1739 was *An Attempt to Introduce the Experimental Method of Reasoning into Moral Subjects*. The model of the 'experimental method of reasoning' that Hume had in mind was undoubtedly Newton's.[27]

But this aspect of Enlightenment Newtonianism did not begin with Hume. It was set in train by Newton himself, who clearly believed that his successes in the physical sciences might well point the way to a new and more certain moral philosophy. He stated this explicitly in the closing words of the *Opticks*:

And if natural Philosophy in all its Parts, by pursuing this Method, shall at length be perfected, the Bounds of Moral Philosophy will be also enlarged. For so far as we can know by natural Philosophy what is the first Cause, what Power he has over us, and what Benefits we receive from him, so far our Duty towards him, as well as that towards one another, will appear to us by the Light of Nature. And no doubt, if the Worship of false Gods had not blinded the Heathen, their moral Philosophy would have gone farther than to the four Cardinal Virtues; and instead of teaching the Transmigration of Souls, and to worship the Sun and Moon, and dead Heroes, they would have taught us to worship our true Author and Benefactor, as their Ancestors did under the Government of Noah and his Sons before they corrupted themselves.[28]

It seems that the earliest attempt to take Newton at his word was by the young Colin Maclaurin in his manuscript 'On the good-seeking forces of the mind' in 1714, but soon more established thinkers would follow suit. George Turnbull (1698–1748), when regent at Marischal College, Aberdeen, published two theses, in 1723 and 1726, which according to his most recent biographer show that 'he was the first Scottish academic to advocate in print the use of the Newtonian method in moral philosophy'.[29] Shortly after this, in 1727, Turnbull resigned his position in Aberdeen, and spent most of the next few years in Edinburgh. His influential *Principles of Moral Philosophy* appeared in 1740, in which he refers to the final words of Newton's *Opticks*, and then says:

It was by this important, comprehensive hint, I was led long ago to apply myself to the study of the human mind in the same way as to that of the human body, or to any other part of Natural Philosophy: that is to try whether due enquiry into moral nature would not soon enable us to account for moral, as the best of Philosophers teaches us to explain natural phenomena.[30]

Another contributor to this movement was Sir John Pringle (1707–1782), later to make a name for himself as a founding father of military medicine. In 1734 he was appointed professor of pneumatics and moral philosophy at Edinburgh, four years after graduating as an MD at Leiden, where he was taught by Boerhaave and the leading Dutch experimental Newtonian physicist, Willem 'sGravesande.[31]

David Hume's *Attempt to Introduce the Experimental Method of Reasoning into Moral Subjects* included a clearly Newton-inspired understanding of the psychological phenomenon of the association of ideas. This is, he wrote, 'a kind of attraction, which in the mental world will be found to have as extraordinary effects as in the natural, and to show itself in as many and as various forms.'[32] Similar ideas appeared subsequently in the *Observations on Man*, of 1749, by the English moral philosopher, David Hartley (1705–1757), but it is quite clear that this kind of moral Newtonianism was already a prominent feature of the Scottish Enlightenment before it was taken up in England. As Roger Emerson has suggested:

> Scots who wished to become the Newtons of the moral sciences could and did borrow a method and an aim from the natural sciences along with the logic which seemed appropriate to them. It was not unnatural that men who had pursued degree courses in which natural philosophy had an importance equal to or greater than moral philosophy should think this way.[33]

This should be enough to substantiate the claim that Newtonianism did not just find its way to Scotland, spreading up from England, nor was it first brought to Scotland by medical students who had been taught their Newtonianism by Boerhaave in Leiden. On the contrary, the Scots were in the vanguard spreading Newtonianism, from Enlightened Edinburgh especially, to other parts of the British Isles and to continental Europe. Furthermore, this was true not just in the physical sciences, but also with regard to the very important Enlightenment development of a supposedly Newtonian moral science. Scots continued to promote Newtonianism throughout the rest of the Enlightenment period, and into the nineteenth century. Newtonianism was very much the dominant approach in the sciences throughout the Enlightenment.

One of the major strengths of Newton's natural philosophy was the fact that it was not yet fixed, and did not claim to have already reached the answers to all questions. In this it provided a major contrast to the Cartesianism which had been enthusiastically embraced in France and the Netherlands. Descartes' system was fully worked out, and was held by its supporters to be capable of answering all questions. As Descartes wrote at the end of his *Principia philosophiae* of 1644 (the final version of his system), 'And thus, by simple enumeration, it is concluded that no phenomena of nature have been omitted by me in this treatise.' To be a Cartesian was to embrace and deploy the already fixed Cartesian system.[34] It was to take French Cartesians some time to see the flaws in Descartes' system, but the backlash set in in the 1730s, when Voltaire (1694–1778) famously declared Descartes' philosophy to be 'no more than an ingenious romance'; Descartes allowed himself to be 'hurried away by that spirit of system which throws a cloud over the minds of the greatest men'.[35]

Newton, by contrast, did not offer a fully worked-out system, but merely a number of suggestions as to where to take things to build on the universally acknowledged triumph of the *Principia*. In the Preface to the *Principia* he pointed out that he had here accounted for the motions of the planets and the tides of our oceans, and he had done so on the assumption of attractive forces operating between bodies. In his own earlier alchemical experiments he had also come to believe in repulsive forces operating between bodies. Accordingly, he wrote that

> we derive from celestial phenomena the gravitational forces by which bodies tend toward the sun and toward the individual planets. Then the motions of the planets, the comets, the moon, and the sea are deduced from these forces by propositions that are also mathematical. If only we could derive the other phenomena of nature from mechanical principles by the same kind of reasoning! For many things lead me to have a suspicion that all phenomena may depend on certain forces by which the particles of bodies, by causes not yet known, either are impelled toward one another and cohere in regular figures, or are repelled from one another and recede. Since these forces are unknown, philosophers have hitherto made trial of Nature in vain. But I hope that the principles set down here will shed some light on either this mode of philosophizing or some truer one.[36]

Newton followed this up in a series of 'Queries' which he added to the end of his *Opticks*. Again, the Query format enabled Newton to make suggestions without committing himself, or his readers, to them.[37] In the first English edition of 1704, and the Latin edition of 1706, the Queries stick to the general suggestion made in the *Principia* Preface and provide suggestions as to how various phenomena, including many chemical phenomena, could be explained on the assumption that the invisibly small particles constituting all bodies interacted by means of attractive and repulsive forces operating between them.

But, responding to criticisms by prominent Continental Cartesians, Newton then added a group of eight extra Queries to the final English edition of the *Opticks* in 1717. Known as the 'Aether Queries', these suggest explanations for phenomena in terms of an all-pervasive aether. These Queries ostensibly look more mechanistic than the other Queries because particles of ordinary matter are seen as completely passive and inert (i.e. devoid of attractive and repulsive forces), and they are moved around by the aether. This is still not compatible with the Cartesian mechanical philosophy, however, because Newton's aether acts on ordinary matter by virtue of strong repulsive forces operating between the particles of the aether, and between the aether particles and particles of ordinary matter.[38]

By 1717, then, Newtonians could choose to explain things on the assumption that all matter is active and endowed with attractive and repulsive forces, or on the assumption that matter is passive and that all its motions are caused by an all-pervasive, active aether.[39] In practice, however, these alternative suggestions are sometimes blended together. For example, recently discovered effects caused by static electricity led to assumptions that electrical spirit was confirmation of Newton's aether. For Edmond Halley, for example, electricity is

> the confirmation of Sir Isaac Newton's notion concerning the existence of an universal medium which he sometimes calls the aether, at other times an electric spirit and which he apprehended was the cause of the phenomena of Gravity, of Light, of Heat, and of Electricity.[40]

Meanwhile, Herman Boerhaave had developed a theory of an all-pervasive fire in the universe, which was responsible for chemical and physiological phenomena. Boerhaave's fire was easily equated with the electric spirit, and again could be seen as equivalent to Newton's aether. But it was also easy for some followers of Newton to suggest that there was an attractive force of gravity, as demonstrated in the *Principia*, and a repulsive force caused by fire – fire is after all auto-diffusive, and causes other bodies to expand. So, here we have Newton's attractive and repulsive forces, but each embodied in different kinds of matter – ordinary gravitating matter, and auto-diffusive, self-expansive fire.[41]

A specific example of this variation of Newtonianism can be seen in Edinburgh in the theorising of James Hutton (1726–1797). In his *Dissertations on Different Subjects in Natural Philosophy* of 1792, he attempted to explain all phenomena in terms of two kinds of matter: gravitating matter, which attracts, and solar matter (so-called because it is seen as originating from the sun), which repels and which is manifested most clearly in light, fire and electricity.[42] Indeed, Hutton's geological theories, published in his *Theory of the Earth* in 1795, are an extended case study in Hutton's Newtonian physics. If gravity were to run unchecked then erosion would eventually result in all land-masses being washed to the bottom of the seas, but the Earth's internal solar matter (as manifested in volcanic eruptions) is continually baking eroded loose matter on the ocean floors into new rocks and then raising those parts of the sea bed to create new land masses; and so the cycle goes ever on, but slowly over vast time scales. 'That Time, which measures everything in our idea, and is often deficient to our schemes,' Hutton famously wrote, 'is to nature endless and as nothing.'[43]

Newton himself, of course, had nothing to say about geology, but there can be no denying that Hutton developed what he saw as Newtonian geology, or a geology based on sound Newtonian principles. Similarly, Newton had nothing to say about how his 'Method' could be used to enlarge 'the Bounds of Moral Philosophy', but it is perfectly clear that Turnbull, Hume and others believed they were developing a moral philosophy based on the application of Newtonian methods. It was in fact the open-ended nature of Newton's hints towards future science that made it such a powerful research tradition (and which enabled it to displace the formerly dominant Cartesianism). It was the possibility for rivalries and contests between different natural philosophers, each laying claim to the best Newtonian credentials, which ensured its fruitful role in the development of science throughout the Enlightenment.

This can be seen, for example, in the development of Newtonian life-science. Pitcairne's and Cheyne's Newtonian iatromechanism clearly still owed a great deal to the earlier iatromechanism of the Cartesians; but the discovery of Trembley's polyp (the freshwater hydra), partheno-genesis in greenflies, and other phenomena which seemed recalcitrantly vitalistic, led to the abandonment of iatromechanism, and a more vitalistic approach to understanding living systems. But the kind of active matter which Newton proposed in the Preface to the *Principia*, and the Queries to the *Opticks*, lent itself very easily to vitalistic conceptions of living systems. In the culminating query, Query 31, Newton spoke of an active principle in matter responsible for 'Fermentation, by which the Heart and Blood of Animals are kept in perpetual Motion and Heat', and without which 'all Putrefaction, Generation, Vegetation and Life would cease'.[44] So Pitcairne's version of Newtonian medicine was dropped, but it was replaced by an equally Newtonian vitalist theory. What is more, this new theory could be embraced freely even by secular or downright irreligious thinkers. While once vitalism went hand in hand with religious belief, after Newton it could be seen as merely an aspect of the active nature of matter.[45]

But this and other aspects of Newtonianism which were seen to have religious or irreligious implications led to yet more contests over the right to call oneself a Newtonian. Thomas Reid (1710–1796) not only opposed Hume's so-called Newtonian moral philosophy, he also rejected the supposedly Newtonian matter theory of Joseph Priestley (1733–1804) on the grounds that it led to irreligious conclusions of which Newton could never have approved. Reid opposed these Newtonians not because he was opposed to Newtonianism, but because he *was* a Newtonian, and didn't want to be associated with atheists, Unitarians or other infidels who falsely claimed to be Newtonian.[46]

Equally, Newtonians could be divided on scientific grounds. Priestley, for his part, saw himself as a Newtonian through and through, and yet he is perhaps most famous for rejecting the new chemistry of Antoine Lavoisier (1743–1794) – but Lavoisier himself was a committed Newtonian. The new concept of chemical affinities was seen in terms of Newtonian forces of attraction and repulsion, and Lavoisier even collaborated with the leading Newtonian mathematician in France, Pierre Simon Laplace (1749–1827), with a view to quantifying theses micro-forces. Writing in his 'Memoir of Affinity' of 1785, Lavoisier mused:

Perhaps one day the precision of the data might be brought to such perfection that the mathematician in his study would be able to calculate any phenomenon of chemical combinations in the same way, so to speak, as he calculates the movement of the heavenly bodies.[47]

The most famous contestants in the history of chemistry in the Enlightenment, Priestley and Lavoisier, were, for all their differences, both equally Newtonian.

Such was Newton's prestige and cachet throughout the Enlightenment, then, that the designation Newtonianism came to be stretched to cover a multitude of positions; and correspondingly, controversies arose when thinkers who claimed to be Newtonian were accused by others of distorting the great man's legacy for their own subversive purposes. It should be clear, anyway, that the stretching of Newtonianism to cover widely different positions is not just strong evidence of how important it was to be a Newtonian in the Enlightenment, but also clear evidence of just how versatile and fruitful Newton's original suggestions proved to be, whether in the science of electricity, of chemistry, geology, or even political economy; small wonder that Newton came to be seen as the embodiment of Enlightenment.

In conclusion, Sir Isaiah Berlin's comment might be reiterated: that Newton's influence was the strongest single factor in shaping the Enlightenment.[48] If that is true, then by showing the prominence of Edinburgh thinkers in the early establishment of Newtonianism, and in its continued promotion throughout the eighteenth century – even to the end of the century, with James Hutton's Newtonian geology – by implication, there is a strong case for the importance of Edinburgh in shaping Enlightenment science, and perhaps even shaping the Enlightenment movement itself more broadly.

Notes and References

1　Hume, D., *The Letters*, ed. J.Y.T. Greig, 2 vols (Oxford: Clarendon Press, 1932), vol. I, p. 255.

2　Smollet, T., *The Expedition of Humphry Clinker*, 3 vols (London, 1785), vol. III, p. 132. The Amyat story appears in Smellie, W., *Literary and Characteristic Lives of Gregory, Kames, Hume, and Smith* (Edinburgh, 1800), pp. 161–2.

3　Hampson, N., *The Enlightenment* (Harmondsworth: Penguin, 1968), p. 16.

4　See Emerson, R.L., 'Science and the Origins and Concerns of the Scottish Enlightenment', *History of Science* 26 (1988), pp. 333–66; and Wood, P., 'Science in the Scottish Enlightenment', in A. Broadie (ed.), *The Cambridge Companion to the Scottish Enlightenment* (Cambridge: Cambridge University Press, 2003), pp. 94–116. Both of these articles ably dispose of those historians who have tried to understand the phenomenon of the Scottish Enlightenment without regard to intellectual developments. See also

Christie, J.R.R., 'The Origins and Development of the Scottish Scientific Community, 1680–1760', *History of Science* 12 (1974), pp. 122–41.

5   Literature on the Scientific Revolution is vast, but see, for example, Dear, P., *Revolutionizing the Sciences* (Basingstoke: Palgrave Macmillan, 2001); and Henry, J., *The Scientific Revolution and the Origins of Modern Science*, 3rd edn (Basingstoke: Palgrave Macmillan, 2008).

6   D'Alembert, J.L., *Preliminary Discourse to the Encyclopaedia of Diderot*, trans. R.N. Schwab (Chicago, IL: University of Chicago Press, 1995), p. 81.

7   Berlin, I. (ed), *The Age of Enlightenment: The Eighteenth-Century Philosophers* (New York: New American Library Inc., 1956), p. 15.

8   This is a historical fact that needs to be explained, but I do not attempt that here – far more research is needed. The answer might lie in differences between England and Scotland with regard to mathematics in education. Certainly, all the early Scottish Newtonians were gifted mathematicians. For a more detailed study of Newton's role in the Scottish Enlightenment, see Wilson, D.B., *Seeking Nature's Logic: Natural Philosophy in the Scottish Enlightenment* (University Park, PA: Pennsylvania State University Press, 2009).

9   Hiscock, W.G. (ed.), *David Gregory, Isaac Newton, and Their Circle: Extracts from David Gregory's Memoranda 1677–1708* (Oxford: for the Editor, 1937); Westfall, R.S., *Never at Rest: A Biography of Isaac Newton* (Cambridge: Cambridge University Press, 1980).

10 Guerrini, A., 'Gregory, David (1659–1708)', *Oxford Dictionary of National Biography* (Oxford: Oxford University Press, 2004); online edn, Jan 2008, http://www.oxforddnb.com/view/article/11456 (last accessed 2 April 2014). It now seems that Edmund Halley was traduced, and that rumours of his atheism were unfair. See Levitin, D., 'Halley and the Eternity of the World Revisited', *Notes and Records of the Royal Society* 67 (2013), pp. 315–29.

11 On Keill, see Schofield, R.E., *Mechanism and Materialism: British Natural Philosophy in an Age of Reason* (Princeton, NJ: Princeton University Press, 1970); Thackray, A., *Atoms and Powers: An Essay on Newtonian Matter-Theory and the Development of Chemistry* (Cambridge, MA: Harvard University Press, 1970); and Hall, A.R., *Philosophers at War: The Quarrel between Newton and Leibniz* (Cambridge: Cambridge University Press, 1980).

12 Keill, J., *Introduction to the True Astronomy* (London, 1721), sig. A3r–v. On Algarotti's *Newtonianismo per le dame* (Naples, 1737), see Mazzotti, M., 'Newton for Ladies: Gentility, Gender, and Radical Culture', *British Journal for the History of Science* 37 (2004), pp. 119–146.

13 Admittedly, much of his fame, or notoriety, stemmed from his role in the calculus dispute between Newton and Leibniz. Arguably, Keill can be seen to have initiated the dispute, and certainly he was known on the Continent as a truculent polemicist. See Hall, *Philosophers at War* and Westfall, *Never at Rest*.

14 Guerrini, A., 'Archibald Pitcairne and Newtonian Medicine', *Medical History* 31 (1987), pp. 70–83; Guerrini, A., 'The Tory Newtonians: Gregory, Pitcairne, and Their Circle', *Journal of British Studies* 25 (1986), pp. 288–311. See also Cunningham, A., 'Sydenham versus Newton: The Edinburgh Fever Dispute of the 1690s between Andrew Brown and Archibald Pitcairne', *Medical History* supplement 1 (1981), pp. 71–98.

15 On iatromechanism, see Frank, R.G., *Harvey and the Oxford Physiologists* (Berkeley, CA: University of California Press, 1980); and Reill, P.H., *Vitalizing Nature in the Enlightenment* (Berkeley, CA: University of California Press, 2005).

16 Lindeboom, G.A., 'Pitcairne's Leyden Interlude Described from the Documents', *Annals of Science* 19 (1963), pp. 273–84. See also Underwood, E.A., *Boerhaave's Men, at Leyden and After* (Edinburgh: Edinburgh University Press, 1977).

17 Lindeboom, G.A., *Herman Boerhaave: The Man and His Work*, 2nd edn (Rotterdam: Erasmus Publishing, 2007).

18 Guerrini, A., 'James Keill, George Cheyne, and Newtonian Physiology, 1690–1740', *Journal of the History of Biology* 18 (1985), pp. 247–66. James Keill was John's brother, and another ardent Edinburgh Newtonian.

19 Henry, J., 'The Reception of Cartesianism', in P. Anstey (ed.), *The Oxford Handbook of British Philosophy in the Seventeenth Century* (Oxford: Oxford University Press, 2013), pp. 116–43.

20 Whiston, W., *Sir Isaac Newton's Mathematick Philosophy more easily Demonstrated* (London, 1716). On Whiston, see Force, J.E., *William Whiston: Honest Newtonian* (Cambridge: Cambridge University Press, 1985).

21 Fletcher, A., *Selected Political Writings and Speeches*, ed. D. Daiches (Edinburgh: Scottish Academic Press, 1979), p. 111.

22 Trail, W., *Account of the Life and Writings of Robert Simson* (London, 1812).

23 Grabiner, J., 'Newton, Maclaurin, and the Authority of Mathematics', in *A Historian Looks Back: The Calculus as Algebra and Selected Writings* (Washington, DC: The Mathematical Association of America, 2010), pp. 229–40, p. 231.

24 Maclaurin, C., *Treatise of Fluxions* (Edinburgh, 1742). The most famous critic of Newtonian calculus was George Berkeley, Bishop of Cloyne, in his *The Analyst: Or, a Discourse addressed to an Infidel Mathematician* (London, 1734). See also Grabiner, J., 'Was Newton's Calculus a Dead End? The Continental Influence of Maclaurin's *Treatise of Fluxions*', *The American Mathematical Monthly* 104 (1997), pp. 393–410; and Sageng, E., 'Colin Maclaurin: A Treatise on Fluxions', in I. Grattan-Guiness (ed.), *Landmark Writings in Western Mathematics, 1640–1940* (Amsterdam: Elsevier, 2005), pp. 143–58.

25 Maclaurin, C., *An Account of Sir Isaac Newton's Philosophical Discoveries* (London, 1748).

26 Campbell, N. and R.M.S. Smellie, *The Royal Society of Edinburgh (1783–1983): The First Two Hundred Years* (Edinburgh: Royal Society of Edinburgh, 1983).

27 Hume, D., *Treatise of Human Nature: Being an Attempt to Introduce the Experimental Method of Reasoning into Moral Subjects* (London, 1739). Capaldi, N., *David Hume: The Newtonian Philosopher* (Boston, MA: Twayne, 1976). On moral Newtonianism, see Emerson, 'Science and the Origins and Concerns of the Scottish Enlightenment'.

28 Newton, I., *Opticks, or A Treatise of the Reflections, Refractions, Inflections and Colours of Light* (New York: Dover, 1979), pp. 405–6. This is a remarkable way to conclude an exemplary textbook of experimental physics (which *Opticks* was), even by eighteenth-century standards. The meaning of the latter part of this passage has only become apparent since the rediscovery of Newton's unfinished attempts to reconstruct what he called *The Philosophical Origins of Gentile Theology* (written and reworked *c.*1680–1700) – on which, see Westfall, *Never at Rest*. The first part, however, proved incredibly influential, and was the stimulus to the science (or sciences) of man on the one hand, and to natural theology on the other.

29 Wood, P., 'Turnbull, George (1698–1748)', *Oxford Dictionary of National Biography*, Oxford University, 2004; online edn, May 2010, http://www.oxforddnb.com/view/article/40216 (last accessed 2 April 2014).

30 Turnbull, G., *Principles of Moral Philosophy: An Enquiry into the Wise and Good Government of the Moral World*, 2 vols (London, 1740), vol. I, p. iii. Turnbull also quotes the influential passage from Newton's *Opticks* on his title page.

31 Emerson, 'Science and the Origins and Concerns of the Scottish Enlightenment', pp. 350–1.

32 Hume, *A Treatise of Human Nature*, Book I, Part I, Section IV.

33 Emerson, 'Science and the Origins and Concerns of the Scottish Enlightenment', p. 349.

34 Descartes, R., *Principia philosophiae* (Amsterdam, 1644), Book IV, Section 199. Henry, 'Reception of Cartesianism'.

35 Voltaire, *Philosophical Letters*, trans. E. Dilworth (New York: Bobbs-Merrill Company Inc., 1961), pp. 64 and 53. See also D'Alembert, *Preliminary Discourse to the Encyclopaedia*, pp. 77–80.

36 Newton, I., *The* Principia: *Mathematical Principles of Natural Philosophy*, trans. I.D. Cohen and A. Whitman (Berkeley, CA: University of California Press, 1999), pp. 382–3.

37 Anstey, P., 'The Methodological Origins of Newton's Queries', *Studies in History and Philosophy of Science, Part B: Studies in History and Philosophy of Modern Physics* 35 (2004), pp. 247–69.

38 For a fuller account see Henry, J., 'Gravity and *De gravitatione*: The Development of Newton's Ideas on Action at a Distance', *Studies in History and Philosophy of Science* 42 (2011), pp. 11–27.

39 See Schofield, *Mechanism and Materialism*; and Thackray, *Atoms and Powers*.

40 Halley, E., 20 May 1731, Bodleian Library MSS Rigaud 37, f. 135. Quoted in Schaffer, S., 'Natural Philosophy and Public Spectacle', *History of Science* 21 (1983), pp. 1–43.

41 Hutton, J., *Dissertations on Different Subjects in Natural Philosophy* (Edinburgh, 1792). Gerstner, P., 'James Hutton's Theory of the Earth and his Theory of Matter', *Isis* 59 (1968), pp. 26–31.

42 Hutton, J., *A Dissertation upon the Philosophy of Light, Heat, and Fire* (Edinburgh, 1794).

43 Hutton, J., *Theory of the Earth*, 2 vols (Edinburgh, 1795), vol. I, Chapter 1, Section 1.

44 Newton, *Opticks*, pp. 399–400.

45 Hall, T.S., *History of General Physiology, 600 BC to 1900 AD*, 2 vols (Chicago, IL: University of Chicago Press); Reill, *Vitalizing Nature in the Enlightenment*.

46 On Reid's dispute with Priestley, see Tapper, A., 'Reid and Priestley on Method and the Mind', in J. Haldane and S. Read (eds), *The Philosophy of Thomas Reid: A Collection of Essays* (Oxford: Blackwell, 2003), pp. 98–112.

47 Lavoisier, A., *Oeuvres*, ed. J.-B. Dumas, E. Grimaux and F.-A. Fouqué, 6 vols (Paris: Impériale, 1862–93), vol. II, p. 550. Priestley's Newtonian credentials are unassailable; see Schofield, R.E., *The Enlightenment of Joseph Priestley* (University Park, PA: Pennsylvania State University Press, 1997); and Schofield, R.E., *The Enlightened Joseph Priestley* (University Park, PA: Pennsylvania State University Press, 2001). Scholarship on Lavoisier tends to emphasise what is new and unique to him, rather than to locate him in the Newtonian tradition, but it is clear that he contributed to Newtonian chemistry when he became aware of the work of Stephen Hales, Joseph Black, Joseph Priestley and other 'pneumatic chemists'. See Guerlac, H., *Lavoisier: The Crucial Year* (Ithaca, NY: Cornell University Press, 1961).

48 Berlin, *The Age of Enlightenment*, p. 15.

THREE

Leiden Chemistry in Edinburgh: Herman Boerhaave, James Crawford and Andrew Plummer

JOHN C. POWERS

The influence of Herman Boerhaave (1668–1738), the illustrious professor of medicine at the University of Leiden, on the early medical faculty of the University of Edinburgh is well documented (Plate 1). According to the standard story, a group of enterprising Scottish students – John Rutherford, John Innes, Andrew St Clair, Andrew Plummer and Alexander Monro *primus* – were all sent to Leiden to study with Boerhaave at some point during the period from 1718 to 1722. After their return to Edinburgh, the first four teamed up in 1724 to offer private courses in medicine and chemistry, which they advertised as following the 'system' of Boerhaave. Soon they began to collaborate with Monro, who taught anatomy at the Surgeon's Hall and then at the university. By February 1726 the five were successful enough to be appointed professors of medicine, thus establishing the Edinburgh medical faculty.[1] A similar story about the importance of Boerhaave for chemistry at Edinburgh has also been told.[2] Plummer did most of the chemistry teaching in the medical faculty, but he was not the first professor of chemistry. James Crawford, who was professor of chemistry and medicine in Edinburgh from 1713 to 1726, holds that honour, but he too earned an MD from Leiden in 1707.

Clearly, having a Leiden medical education was an important credential in eighteenth-century Edinburgh, and by the 1720s Boerhaave was an extremely influential name in medical circles. Yet one might consider whether these early Edinburgh professors simply invoked Boerhaave's name, or actually followed Boerhaave's system in medicine and chemistry. Scottish students had been travelling to Leiden to study medicine for decades before Boerhaave joined the medical faculty. This suggests that training in Leiden was seen as an important credential for a physician long before Boerhaave began his career. Ironically, there have been few studies that analyse the extent to which Boerhaave's ideas and methods specifically appeared in Edinburgh courses. One reason for this lack of analysis may be because Boerhaave, even in his own time, was seen as a pedagogue and systematiser. In the colourful words of Scottish physician and Leiden graduate James Houston, the Dutch professor was skilled at 'digesting a huge heap of jargon and indigested stuff into an intelligible, regular, rational system.'[3] This situation, however, has proven to be an enormous obstacle for historians who have attempted to find the philosophical core of Boerhaave's system, proposing empiricism, mechanism, Calvinism and Newtonianism as overarching frameworks for his work. Each of these studies has made a contribution, but as a group they present a contradictory and unsatisfying account.[4] By contrast, important questions about Boerhaave's aims and approach can be answered by examining the structure of his system or, rather, the pedagogical method (i.e., ordering) of his courses and how he put them together. He styled himself as a reformer, and so his courses, especially his chemistry courses, did not replicate the standard courses and textbooks of his day. Whereas most chemistry courses focused on recipes to make things, Boerhaave was much more interested in theory; that is, understanding and categorising chemical phenomena and the natural principles behind them. Therefore he addressed a wider range of topics, utilised different taxonomic categories and presented a more rigorous philosophical approach than that found in craft-oriented textbooks.

In this chapter Boerhaave's chemistry, as presented in his courses at Leiden and in his textbook *Elementa chemiae* (1732), is compared with that of the first two professors of chemistry at Edinburgh, James Crawford and Andrew Plummer. First, a review of the salient features of Boerhaave's chemical work is provided, which discusses its origins within the Leiden medical faculty and Boerhaave's attempts to integrate it more fully into the medical faculty by making it more theoretical. Then Crawford's and Plummer's work is examined within the context of establishing their academic medical careers and interacting with the local scholarly community. For Crawford, an extant set of student notes from his first chemistry course shows that he followed the same pedagogical structure and theoretical concepts found in Boerhaave's contemporary courses. No complete set of course notes seems to exist for Plummer, so two papers that he read at the Philosophical Society of Edinburgh, which reveal the influence of Boerhaave's teachings, are analysed. Ultimately, both of these men utilised Boerhaave's chemistry as a resource and foundation for their own chemical courses and work. Yet both also progressed beyond this foundation in adapting their work to fit the local circumstances and addressing innovations in the field.

Boerhaave's Reform of Chemistry in Leiden

Before the 1720s, Scottish medical students who could afford to study on the continent inevitably gravitated to Leiden because of the reputation and resources of the university's medical faculty. By the middle of the seventeenth century, Leiden boasted a first-rate botanical garden, an anatomy theatre, natural history collections and two wards in a local hospital set aside for clinical instruction.[5] Medical students in Leiden reaped the benefits of these resources and gained a great deal more practical experience than they would have at other centres of medical education. For example, the Scottish physician Robert Sibbald (1641–1722) reported that when he was a student in 1660 and 1661, he witnessed 23 dissections and autopsies.[6] In addition, the university curators and town burgemeesters endeavoured to attract 'illustrious' professors, who tended to have progressive philosophical views on medicine, to the medical faculty. Leiden was one of the first schools to debate and teach William Harvey's theories regarding the circulation of the blood and also the philosophy and medical theories of René Descartes.[7] Finally, students who came to study in Leiden enjoyed a level of religious toleration which was typical of public life in the Dutch Republic, but not elsewhere in Europe. Although professors were required to be members of the Dutch Reformed Church (Calvinist), there are records of students from all Protestant faiths and even a few Roman Catholics. In Boerhaave's day there was at least one Greek Orthodox student from Constantinople.[8]

The most distinguished professor in the 1660s, and Sibbald's mentor, was Franciscus Sylvius (François de le Boë, 1614–1672), who had become famous for his iatrochemical approach to medicine. Sylvius theorised that physiological processes, notably digestion and metabolism, were controlled by the reactions of acids and alkalis in the body. An excess of an acid or alkali caused disease, which could be remedied by re-establishing the proper balance through medicaments containing ingredients with the opposite nature.[9] After his death, however, his colleagues at Leiden's medical faculty quickly abandoned Sylvius's theory of disease for being too simplistic; even Boerhaave criticised this approach in a 1718 university oration.[10] Nevertheless, Sylvius popularised the use of chemical analysis in anatomical research and chemical concepts in theories of physiology. He produced a group of students who carried on his anatomical work, and his successors on the medical faculty, including Boerhaave, all performed basic chemical analyses on bodily fluids and tissues as a regular part of their anatomical practice.[11]

Sylvius's most important contribution to chemistry at Leiden was his ongoing effort to found a chemical laboratory and chair of chemistry. Because chemistry was a central part of his medical programme, Sylvius probably taught private chemistry courses during the 1660s. By the end of the decade, his student Lucas Schacht (1634–1689) had taken over this teaching, which focused on pharmacy and *materia medica*.[12] In 1669, after an extended campaign during which Sylvius threatened to resign, the Leiden curators agreed to establish a chair in chemistry. They appointed Carel De Maets (1640–1690), who had a degree in philosophy from Utrecht and had worked in the laboratory of the well-known chemist Johann Glauber (1604–1670) in Amsterdam.[13] This position, however, was tenuous. De Maets received no salary, and after Sylvius's death in 1672, he was shuffled between the philosophy and medical faculties. Only in 1679 was he made a professor *ordinarius* on the medical faculty, primarily so he could teach a medical *praxis* course. In addition, his chemistry courses had competition from two extra-mural lecturers, Jacob Le Mort (1650–1718) and Christiaan Marggraf (1626–1687), whose competing views on chemistry further challenged De Maets's status. Leiden students attended all three courses and freely compared and combined their notes, as demonstrated when an English student, Christopher Love Morley, published his notes as a textbook, *Collectania Chymica Lydensia* (1684).[14] As yet another indicator of the low status of the professor of chemistry, when De Maets died in 1690, Le Mort's petition to replace him was blocked by the medical faculty on the grounds that it would tarnish the 'lustre' of the university. Le Mort was made '*praefectus*' of the university laboratory, but was forced to wait for 12 years before being named chair of chemistry.[15]

Boerhaave studied medicine and chemistry in Leiden during this period, when there was no professor of chemistry. The son of a local Calvinist minister, he was a student in the arts faculty at Leiden from 1684 to 1690 with the intention of studying theology. At the suggestion of one of his theological mentors, he began to study medicine too, but he never matriculated in medicine at Leiden. Rather, he attended the public anatomies of Charles Drélincourt (1633–1697) and public and private anatomical demonstrations of Anton Nuck (1650–1692). He obtained his theoretical education in medicine and chemistry by reading all the relevant books held in the university library, in which he worked from 1691 to 1693. As he later related, he began with the ancient authors and proceeded chronologically through the most recent works.[16] Since the chair of chemistry was vacant, and Le Mort taught only sporadically, Boerhaave apprenticed himself to a local apothecary, David Stam (1633–1711), who had earlier worked with Sylvius and who trained Boerhaave how to perform chemical operations.[17] In July 1693 Boerhaave travelled with his completed medical thesis in hand to the Guelders Academy in Harderwijk, where the graduation fees were much lower than in Leiden, and received his medical degree after three days of examination.[18] After a confrontation on a canal boat, which led to a rumour that he was a Spinozist and undermined his chances for the ministry, Boerhaave returned to Leiden to establish a private medical practice.[19] According to his own account, he set up a small chemical laboratory in a shed at his house.[20]

In 1701 Boerhaave and his patron, Johannes van den Berg (1664–1755), the secretary of the university curators, conspired to take advantage of a temporary monetary and manpower problem at the university. Faced with a shortage of medical professors, the curators appointed Boerhaave as a lecturer on the institutes of medicine, the basic medical theory course. Within a year, a group of 'foreign students' petitioned the curators to allow Boerhaave to offer courses in chemistry as well.[21] He began his first chemistry lecture course shortly thereafter, followed by the same lecture course with a practical, operations course in the winter of 1702–3. In May 1702, the fortunes of the medical faculty changed drastically when King William III, who was also the Dutch Stadholder and self-proclaimed 'opper-curator' of the university, died after a fall from his horse. This event led to the university receiving an influx of money from the Dutch state, and the appointment of two new medical chairs, including Le Mort as chair of chemistry.[22] Despite this new situation, Boerhaave was permitted to continue teaching chemistry and medicine due to his popularity with students. When in 1703 he was offered a chair in medicine at Groningen, the Leiden curators, citing 'the influx of native and foreign students' to his courses, promised him the next chair that should become available. Boerhaave was appointed chair of botany and medicine in 1709.[23]

During his career at Leiden (1702–38), Boerhaave completely reformed the way that chemistry was taught. As a physician who followed the Leiden tradition of utilising chemical methods in anatomy and physiology, he shaped his chemistry courses in a way that would allow them to be integrated more fully into the medical curriculum. As a result he placed a great deal more emphasis on chemical theory, so that his courses would follow the standard pedagogical order of medical courses: the institutes (theory) course followed by the praxis (application of theory) course.[24] Most chemistry courses, such as those given at Leiden by De Maets, Le Mort and Margraff, focused predominately on recipes for chemical products; primarily medicaments, but also perfumes, cosmetics, soaps, etc. This emphasis was evident in the published versions of these notes, *Collectanea Chymica Leydensia*, which begins with an 18-page theory section presented as a prologue, followed by 480 pages of recipes presented in alphabetical order.[25] This was the common structure; Nicolas Lemery's *Cours de chymie*, the most popular textbook of the day, included a 68-page 'prelude' defining operations, equipment and principles, then 880 pages of operations in the 1713 edition.[26] By contrast, Boerhaave presented an entire lecture course on chemical theory, and then a second course on chemical operations. When he published his textbook *Elementa chemiae* (1732), the first volume on the history and theory of chemistry comprised almost 900 pages in Latin, while the second volume on operations comprised just under 600 pages.[27]

Boerhaave's emphasis on chemical theory demanded a more rigorously philosophical approach than he found in extant textbooks and courses. The theory section of most didactic courses contained a discussion of some system of chemical principles. De Maets, for example, advocated the *tria prima* of Paracelsus – salt, sulphur and mercury – while Lemery proposed five principles: salt, sulphur, mercury, earth and phlegm (essentially, water). Each of these principles conveyed specific chemical and empirical properties to the substances in which they inhered: sulphur conveyed inflammability, colour and odour; mercury conveyed fluidity and metallic properties; etc. Lemery claimed that these principles were the constituents of substances and could be separated from bodies though chemical operations such as distillation or combustion.[28] Boerhaave, however, rejected the various systems of chemical principles based on a critique of these entities that had developed during the second half of the seventeenth century. Several prominent chemists, most notably Robert

Boyle (1627–1691), contended that the substances which the 'chymists' identified as principles or elements were in fact compound bodies generated by the chemists' operations themselves. In other words, they were artefacts of the chemist's fire.[29]

Boyle's arguments induced Boerhaave to adopt a new approach to the theory of chemistry, which centred on the so-called chemical instruments. The instruments (fire, air, water, earth, *menstrua* and chemical vessels) were the tools, natural and artificial, that the chemist manipulated during chemical operations. The instruments were a heterogeneous group, both ontologically and functionally. 'Fire' was an imponderable, particulate fluid, which acted as the cause of phenomena relating to heat, such as the expansion of a body's volume, flame and the destructive dissolution of bodies. 'Air' was a ponderable fluid which exerted pressure on bodies necessary, for example, during combustion, and also acted as a medium to contain vapors and 'spirits'. 'Water' was a ponderable fluid and the most common solvent found in the chemist's laboratory. 'Earth' was a simple, hard, insoluble body which remained fixed in the fire, and whose function as an instrument was to fix volatile spirits, salts and oils. Finally, the term *menstruum* did not refer to a specific species of body, but generally to the power of chemical species, such as acids, oils and alcohol, to act as solvents.[30]

The instrument approach to chemistry shifted the focus of chemical theory in Boerhaave's courses on to the effects that chemical operations, as conducted via the instruments, had on matter. Boerhaave's source for the instrument theory was Johannes Bohn (1640–1718), a professor of medicine at Leipzig who both endorsed Boyle's critique of the chemical principles and offered the instruments as an alternative approach. Bohn's *Dissertationes Chemico-physicae* (1685) organised the instruments and their uses into a series of topical dissertations and, as such, provided Boerhaave with a pedagogical model for his first chemistry course.[31] In Boerhaave's courses, the instruments functioned as taxonomic categories for ordering and explaining relevant chemical and physical phenomena. For example, when discussing the instrument fire, Boerhaave defined two types: 'shining' (*lucens*) and 'burning' (*urens*). Under the category of shining fire he described the action of burning lenses and mirrors, which were then a topic of research for chemists at the French Académie Royale des Sciences.[32] He divided burning fire into several sub-categories, including the generation of heat through friction or concussion (such as striking a metal plate with a hammer), strong reactions of chemical species (such as acids and alkalis), and combustion.[33] By acting as pedagogical *loci*, the instruments played a central role in the structure of Boerhaave's course. At the start of his course he discussed the 'objects' of chemistry – the natural substances and simples upon which chemical operations were performed – then he proceeded to the instruments before concluding with operations, which examined specific techniques for utilising instruments to effect change in objects.[34]

Boerhaave adapted the instrument theory to both the empirical philosophy and corpuscular theory of matter that he embraced. Throughout his medical and chemical work, he often devised explanations for phenomena in terms of the shapes, textures and motions of corpuscles. However, he always devised these explanations after the fact and never used particle shape and motion to predict phenomena. In his first academic oration at Leiden (1701), he railed against Cartesian physicians, who utilised their mechanical models in this speculative manner, and suggested that the proper method in medicine was built upon the collection and study of many instances before a phenomenon could be understood properly.[35] Boerhaave adopted this method in chemistry as well, arguing that chemical properties – those not related to extension, solidity, weight, etc. – could only be determined experimentally. To generate this kind of knowledge, he advocated systematic experimentation and pointed to Francis Bacon as his model. Each chemical species should be mixed with others and under different conditions to reveal the latent properties, which emerged and could not be revealed in any other way.[36]

Through these systematic experiments, Boerhaave created what he called 'histories' of chemical phenomena, usually built around specific instruments. For example, in the *Elementa* he described a long series of experiments, in which he mixed together various solvents and salt solutions and measured the change in heat using a Fahrenheit thermometer. He referred to these experiments as a component in creating the 'certain history of heat'.[37] The two aspects of Boerhaave's approach were most evident in his discussion of chemical *menstrua*. The main phenomenon to be explained was the differential action of solvents; why a specific solvent dissolved some types of bodies, but not others. In his first course, Boerhaave provided two taxonomies for *menstrua* reflecting his two approaches. The first examined the theoretical causes of the action of solvents on bodies based on particle shape and motion: whether the points of the solvent particles matched the pores of the solute, the cohesion of the solute particles, and the motion of the solvent particles.[38] The second listed species of chemical solvents – water, oils, alcohol, acids and alkalis – and their empirical, chemical properties.[39] Nevertheless, he argued that one could not predict the effect of a specific *menstruum* on a specific species of body until one tested the interaction experimentally.[40]

When Boerhaave was appointed to the chair of chemistry in March of 1718 he was obligated to offer one public course on chemistry, which was available to all matriculated students without lecture fees. In October he began an extended course on chemical instruments, presenting an exhaustive account of each instrument in turn over a period of several years.[41] The 'instruments course' revolutionised the teaching of chemistry because where he could, he presented important claims about the properties of each instrument through demonstration-experiments. He structured these demonstrations according to what he called the 'method of the geometers', indicating that his manner of presenting the demonstration and its interpretation derived from practical, mathematical sources such as Newton's *Opticks*.[42] For each demonstration, he performed the experiment, described the results and then interpreted the significance of the experiment through numbered 'corollaries'. The aim of these experiments was to reinforce the empirical foundations of chemistry by showing that the precepts which guided chemical practice derived from natural phenomena, not speculative theories. Logically ordered sequences of these demonstrations constituted 'experimental histories', from which theoretical claims could be drawn. As such, these demonstrations conflated the distinction between research and pedagogy. In fact, Boerhaave utilised this 'geometrical' structure to present demonstration-experiments in the *Elementa chemiae*, but he also used the same structure to present his research on the transmutation of mercury, which he published as a series of three papers in the *Philosophical Transactions of the Royal Society of London* (1733–6).[43]

When Boerhaave began sending friends and colleagues copies in the autumn of 1731, the *Elementa chemiae* was the culmination of almost 30 years of chemistry teaching and experimentation. The textbook combined his three courses: his term-length lecture and operations courses, and the instruments course. Whereas his lecture notes only provided an outline of the main points of his courses, the *Elementa* presents the mature version of his thinking on chemistry, wandering off topic at times, perhaps mimicking Boerhaave's presentation in the lecture hall. The book was a new kind of textbook in that it was written for novices but focused, like the instruments course, on making knowledge as well as making things. Many of Boerhaave's discussions and demonstration-experiments utilised new apparatus in chemistry, such as thermometers, air-pumps and burning lenses, which had little or no presence in the craft chemistry of prior textbooks. These apparatus were philosophical instruments, meaning that they generated or measured new phenomena, but were not used to make anything.[44] In addition to providing students with a model of how to create chemical knowledge via experiment, Boerhaave encouraged them to go out and do it, in effect, identifying research problems within his chemical system. He challenged his readers and, one presumes, his students in his courses to continue his investigations by, for example, obtaining a Fahrenheit thermometer and completing the 'certain history of heat', or searching for 'the hidden virtue' in the air, which endowed it with life-sustaining properties.[45] Even the operations section, based on his term-length course, was not organised as a recipe book but rather as a pedagogical manual, proceeding from the most simple to the most complex operations. Boerhaave began with what was in his mind the simplest and most fundamental operation: the extraction of essential oils from plant matter. He then progressed to distillation and other processes related to substances from plants, thence to animal substances, and finally to minerals.[46]

Boerhaave's chemistry was extremely influential. The *Elementa* was issued in at least forty separate printings from 1732 to 1791, not including several abridgements, and was translated from Latin into English, French, German and Russian.[47] The *Elementa*, however, came relatively late in Boerhaave's career; the founders of the Edinburgh medical school had all left Leiden at least ten years before the book was published. Prior to the 1730s, Boerhaave's students disseminated his ideas and methods. During his 37-year career (1701–38), more than 1,900 students from across Europe matriculated at Leiden, including 205 Scots.[48] Many of these young men, carrying their lecture notes and experiences with them, returned to their home countries and established medical teaching based on the Leiden model.[49] Since chemistry was an integral part of Boerhaave's method of medical education, the institutions which adopted his system tended to offer chemical instruction as well. This was the case for the University of Edinburgh.

James Crawford and the First Chemistry Courses in Edinburgh

As in Leiden, chemistry teaching in Edinburgh was intimately connected to the rise of the university medical school. By the end of the seventeenth century, the town had become a centre for medical learning and practice in Scotland. The activities of professional guilds, such as the Incorporation of Surgeons (which included apothecaries from 1657) and the recently founded (1681) Royal College of Physicians of Edinburgh, created a demand for instruction in medicine, botany, anatomy and chemistry due to all of the medical and surgical apprentices in town.[50] In 1685 the Town Council attempted to create a *de facto* medical faculty at the university by appointing three professors of medicine: Robert Sibbald (1641–1722), James Halket (c.1652–1711) and Archibald Pitcairne (1652–1713). Unfortunately, these appointments came to nothing, as each man was forced out of his post in the political turmoil of the late 1680s.[51] However, independent lecturers, who were typically supported by the surgeons, physicians, or both, increasingly taught courses on anatomy, botany and, on occasion, chemistry. James Sutherland (1638–1719) was appointed Intendant of the town's physic garden in 1676 and commissioned to teach botany to all apprentices in surgery and medicine.[52] In 1697 the Incorporation of Surgeons constructed an anatomical theatre, which included a chemistry laboratory. In addition to yearly dissections, Alexander Monteith leased the laboratory space and offered at least one documented chemistry course in 1702.[53] In 1705, the Incorporation and Town Council named Robert Eliot, who had briefly been a student of Pitcairne's at Leiden (1691–2), 'Public Dissector of Anatomy', a post which required him to give regular anatomy courses and maintain the university's collection of 'rarities'.[54] For medical courses proper, college members such as Robert Sibbald taught private courses on the institutes (theory) and practice of medicine; and lecture notes, notably those of Archibald Pitcairne, circulated among students.[55]

By the early eighteenth century, professorships at the university also became available to ambitious instructors. As Roger Emerson has pointed out, all university appointments in the eighteenth century were reliant upon patronage, and candidates had to build coalitions of patrons to secure their positions. For medical appointments, a candidate first had to secure the surgeons and physicians, who determined with which instructors their apprentices would study. But ultimately the candidate had to court powerful politicians, who influenced the Edinburgh Town Council, which had the legal right to make university appointments.[56]

Take, for example, the teaching of botany. At the behest of the surgeons and physicians, James Sutherland was appointed professor of botany for the university in 1695. When in 1705 the Incorporation of Surgeons determined that Sutherland was no longer fulfilling his duties properly, they began sending their apprentices to Charles Preston (1660–1711), a well-connected virtuoso with a Leiden medical degree. Sutherland resigned the following year, and Preston was appointed chair of botany.[57] This series of events typifies the pattern for early Edinburgh medical appointments: one needed the support of the Incorporation or College, and increasingly a Leiden degree or, at least, time spent in Leiden. The latter requirement was to satisfy the members of the Edinburgh College of Physicians, several of whom (Sibbald, Halket) had taken their medical degrees at Leiden, and Pitcairne had taught there briefly from 1692 to 1693.[58] In addition, the Incorporation of Surgeons also boasted Leiden graduates, notably John Monro (1670–1740), who was influential in founding the medical faculty.[59] These men saw the Leiden curriculum as the model for proper medical education and used their influence to shape the medical institutions in Edinburgh accordingly. Upon his return from Leiden in 1693, Pitcairne revised the College of Physicians' license examination to make it more similar to the Leiden medical faculty's graduation examination.[60]

James Crawford's appointment as professor of chemistry and medicine can be seen as a study in building a career in medicine and academia in Edinburgh. Born in Leith, he may have studied for a time in Edinburgh and then travelled abroad, although little is known about his specific training. He matriculated into the medical faculty at the University of Leiden on 31 May 1707, and successfully defended his thesis, 'De Scorbuto' (On Scurvy) on 6 July, graduating after only five weeks. In 1708 he was awarded a second MD *ad eundem* from King's College, Aberdeen, with the support of Archibald Pitcairne. By November 1710 he was in Edinburgh and applied to take the medical licensing examination administered by the Royal College of Physicians, which he passed. In February 1711 he was elected a fellow of the College, and on 6 December of that year he was elected secretary and librarian.[61] At this point, Crawford had garnered the proper credentials and seemingly secured the support of the Edinburgh medical establishment, placing him in a good position for a university appointment. He got his opportunity in early 1713, as news spread that the University of Glasgow had received permission to found a chair of medicine. As soon as he returned to Edinburgh from a trip that kept him out of town for most of the summer, Crawford petitioned the university's principal, William Castares (1649–1715), to create a similar chair for him in Edinburgh. Much of the subsequent negotiation and politicking to secure this position is lost to the historical record, but Roger Emerson has suggested that Crawford and Castares secured the support of James Graham, Duke of Montrose (1682–1742), who would soon occupy several high government posts, including the Lord of the Regency and Scottish Secretary, as well as Chancellor of the University.[62] In November of 1713, Castares sent a letter to the Royal College of Physicians, asking for their endorsement of Crawford. Once this was obtained, Castares then proposed Crawford to the Town Council in December, citing the 'want of professors of physick and chemistry' in Scotland. The Council appointed Crawford on 9 December without salary and granted him access to two rooms for teaching.[63]

Crawford's appointment is commemorated in this volume as the first chair of chemistry at the University of Edinburgh, but there is no evidence to suggest that he intended to found a programme in chemistry. For Crawford, the post may have been a mere stepping stone for establishing his academic career. In January 1714, he sought and received permission from the Town Council to apply for the chair of medicine at Glasgow, a position which included a £50 per annum salary. He did not accept this appointment. In a broadside he published in 1719, he claimed that he turned down Glasgow's offer to preserve the reputation of Edinburgh. He made this claim, however, within the context of campaigning for Edinburgh's vacant chair of Hebrew, to which he was appointed and for which he received a £100 per annum salary.[64] As for his appointment as professor of chemistry, Castares' letter to the College of Physicians did not mention chemistry as constituting part of the position.[65] It is likely that Castares added the focus on chemistry later as a means of justifying the chair's creation, since instruction in botany and anatomy were already available in Edinburgh. Crawford taught only a few chemistry courses in his 12 years of holding the chair, probably just enough to keep his patrons satisfied, and there is no record of him teaching a single medical course. He certainly collected lecture fees from students for his chemistry courses and, beginning in 1718, fees for medical degree examinations, which he conducted with assistance from his colleagues at the College of Physicians. It is unclear how successful Crawford was in private medical practice, but the status he gained as university professor and fellow of the Royal College would have helped attract patients. An account of a case in 1721 published by Joseph Gibson, an Edinburgh surgeon, names Crawford as a consulting physician who assisted in a dissection. At this same time, he assisted in revising the official Edinburgh Pharmacopeia.[66] After 1721, he seems to have spent much less time on chemistry and medicine, shifting his efforts to Hebrew and other humanistic pursuits.

Nevertheless, during his first few years as chair, Crawford did teach a few chemistry courses. Two sets of notes from his courses exist, each of them dating from the period of 1714–19.[67] There is circumstantial evidence that he performed demonstrations in his courses, since in January 1715 the Town Council granted him £10 for equipment.[68] His last course for which there is any evidence began in November 1720, as advertised in the September *Caledonian Mercury*.[69]

Based on extant student lecture notes, when he did teach, his chemistry courses were modelled on those of Herman Boerhaave. The notes examined here derive from Crawford's first course (c.1714), taken by John Fullerton, and are compared to Boerhaave's own course notes.[70] The structure of Crawford's course, the chemical theories he discussed and his general notions regarding the state of chemistry as an academic field are almost identical to those found in the term-length lecture course that Boerhaave routinely gave at Leiden from 1702 to 1717. Both professors' courses, after a preliminary discussion, presented a brief history of chemistry followed by a section on chemical theory, which examined the 'objects' (common substances or simples) and 'instruments' of chemistry. The last half of Crawford's course, in which he discussed and demonstrated common chemical operations, mirrored Boerhaave's operations course. There is evidence to suggest that Crawford attended one of Boerhaave's courses, perhaps in 1706 or 1707, before he matriculated at Leiden, or that he obtained a detailed copy of the course notes to use as a model.[71] Crawford presented the same theories, and many of the same examples and evidence, found in Boerhaave's Leiden courses. For instance, when he discussed the specific weights of metals, he quoted the values that Boerhaave reported in his early courses – which he then appended with measurements 'lately tried' in Edinburgh.[72]

Like Boerhaave's course, Crawford's presented a call for the reform of chemistry. He began by lamenting that chemistry as commonly taught 'is ye most confused and obscure of all sciences'. His aim was to show the reason for this obscurity, and detail what methods could be utilised to 'revise' chemistry, 'that it may deserve its place amongst ye sciences'.[73] The main problem was that chemistry is an experimental science, but many chemists, who were 'dexterous' at making experiments, were unlearned and could not reduce their practical experiments 'into ye method of a science'. Thus, Crawford advocated that chemistry must be reformed by learned men, but not by those who wished to apply a philosophical system or 'Physical Hypothesis' to the art. Rather, these reformers had to proceed empirically, so that they would 'understand the real properties of bodies, the powers and forces, from which they could deduce the causes of chemical appearances'.[74] Crawford suggested that chemical experiments be made in 'geometrical order' from simple to complex. First, simple chemical bodies would be applied to other bodies and to each instrument, and then the products of these experiments would be applied to new bodies until all the various combinations of products were known. Chemical doctrines would only be established *a posteriori*.[75] This method was built into the structure of Crawford's course, especially in the second half, where he described chemical operations. Following the order of Boerhaave's courses, the first operation discussed was the extraction of an essential oil from plant matter. From there the operations became more complex, moving to plant distillation and the synthesis of basic herbal medicaments, then to animal simples, and finally concluding with operations on minerals.[76]

Crawford began the theory section of his course with a discussion of the 'objects' of chemistry, which included an examination of the elements or principles of matter. Following Boerhaave, Crawford divided the objects into taxonomic categories – vegetable, animal and fossil (i.e. mineral) – providing descriptions of common chemical simples in each category.[77] He then began his discussion of chemical principles by asking, 'how far [does] chymie go in the solution of bodies?'[78] Like Boerhaave, he adopted Robert Boyle's scepticism regarding the chemical principles and peripatetic elements. He rejected both as 'false and groundless', and he concluded based on arguments cribbed from Boyle and Boerhaave that 'the Principles Chymists extract from bodies want ye simplicity of elements and are rather mixed bodies'.[79] In place of the principles, Crawford accepted a moderate form of atomism, in which chemical properties and change were explained via the size, figure and motion of particles. He was careful not to apply his particulate 'hypothesis' as a predictor of chemical action or causes: one should not 'vent fictitious hypotheses about ye figure, size of ye partes of ye bodies, or ye pores' without experimental proof, for 'it does not cure our ignorance to pretend knowledge'.[80] Nevertheless, he provided his students with a list of commonly-accepted chemical principles (salt, water, oil or sulphur, and earth) which were not elements, but rather categories of products from distillation and other forms of chemical analysis.[81] So, under the heading of each of these principles, save water, Crawford enumerated and described the specific kinds of salts, oils and earths that a chemist encountered. But, he reiterated, the properties of these substances, through which they are identified, cannot be predicted by any mechanical or corpuscular theory. As he stated regarding salts: 'The particles of salt, not falling under our senses, we cannot know anything of their composition or mechanical character *a priori*, only by their effects we are sure they must be rigid and of a penetrating nature.'[82]

The core of Crawford's chemical theory was the chemical instruments, which brought chemical species into motion and through which the chemist effected his operations. Crawford's instruments were Boerhaave's instruments – fire, air, water, earth, *menstrua* and chemical vessels. This section of Crawford's course was identical in structure to Boerhaave's. For example, in his presentation of chemical *menstrua*, he provided a definition of chemical *menstruum*, discussed their mode of action, and then outlined the various kinds of *menstrua*: watery, oily or sulfurous, alkaline and acidic, finally concluding the section with comments on the alkahest, or universal solvent. In his course, Boerhaave presented chemical *menstrua* following exactly the same order of topics.[83] In addition, Crawford adopted Boerhaave's theory of the action of *menstrua*, which held that solvents worked through mechanical action. In this theory, the particles of the *menstruum* insinuated themselves into the pores of the dissolvent to break it apart. The most important phenomenon to be explained by the theory was the selective action of *menstrua* – how and why a specific *menstruum* acts on some bodies, but not others. Crawford followed Boerhaave in presenting three 'principles' for this differential action based on the size, shape and rigidity of solvent and solute particles and their 'active parts'.[84] While operating within Boerhaave's structure,

Crawford shaped his lectures to accommodate his local circumstances. At the start of his presentation of *menstrua*, he reduced what was in Boerhaave's course a much longer discussion of the mechanical action of *menstrua* into four 'rules', in the first of which he attributed his mechanical theory to Robert Boyle.[85] Crawford probably made this abridgement to accommodate his time restraints, and the attribution to Boyle reflected the fact that Boyle was probably much better known in Edinburgh at the time than Boerhaave, who was still establishing his reputation.

Crawford challenged other key ideas in Boerhaave's chemistry more forcefully, even while he maintained the Leiden professor's method. In his lectures on the instrument fire, Crawford attempted to combine Boerhaave's subtle fluid with several of Isaac Newton's assertions on heat and light found in the first two editions of the *Opticks* (1704; 2nd Latin edition, 1706). For example, he categorised two varieties of fire, 'celestial' and 'artificial', which he described utilising a topical structure that was parallel to Boerhaave's 'shining' and 'burning' fire, but which deviated from Boerhaave's understanding. Crawford argued that light was celestial fire, which once it came into contact with objects on earth as solar rays, remained as Boerhaave's subtle fluid. Boerhaave, however, was ambiguous about the relationship between fire and light, stating only that sunlight caused fire to move parallel with it.[86] Crawford's stronger stance derived from his support of Newton's theory that light was composed of material particles, which can be bent by mirrors, lenses, thin films and crystals, such as the famous Iceland spar.[87] This theory, however, was easily reconcilable with Boerhaave's and integrated into the Boerhaavian structure of Crawford's course. Crawford's section titled 'celestial fire' examined the ability of burning mirrors and lenses to 'collect fire' at their foci, just like Boerhaave's section on 'shining fire'. Both courses also discussed the same examples, the burning mirror of François Villette (1621–1698) and the Tschirnhaus lens used by Wilhelm Homberg (1652–1715) at the Académie Royale des Sciences.[88] In other cases, however, Crawford rejected Boerhaave's explanation of phenomena. In his discussion of 'artificial' fire (i.e. flame), he challenged the 'opinions' of 'Bernoulli, Boerhaave, and Cheyne' that fluid fire needed 'sulpherous matter' to ignite into flame. Instead, he offered Newton's theory of flame: that heat consisted of the vibrating parts of bodies, and when the vibrating motion exceeded a certain degree, it caused the body to glow and become rarefied into smoke, thus producing flame.[89]

But Crawford did not accept the Newtonian programme uncritically either. By the 1710s, English and Scottish physicians who were supporters of Isaac Newton, such as George Cheyne (1671–1743), James Keill (1671–1721) and William Cockburn (1669–1739), were attempting to integrate Newtonian concepts of force and methods of mathematical analysis into medicine. In Edinburgh, Pitcairne, who had taught several of these physicians, was the leading advocate for this approach until his death in 1713.[90] Some of these men, notably Keill and John Freind (1675–1728) endeavoured to apply Newtonian philosophy to chemistry as well. Freind gave a series of lectures on chemistry at Oxford in 1704 based on a chemistry manuscript from Newton, *De natura acidorum*, which had been circulating among Newton's supporters. In his lectures, Freind stated that he intended to reduce chemistry to 'the roots of true philosophy', by which he meant Newtonian forces of attraction and repulsion. Following Newton's 'geometrical method', he began the course with a list of postulates in which he described the attractive forces between particles of matter, and how these forces varied according to the shape, texture and density of the particles.[91] Crawford, however, rejected this approach to chemistry on the grounds that it was a 'philosophical hypothesis' not grounded in experimental evidence. He argued that the attraction of bodies based on the inverse square of their distance may be useful for visible bodies, but asked: how could this notion be used for unobservable particles without 'assuming at pleasure' the 'figure, distance, and solidity of their parts'?[92] Here, Crawford was protecting Boerhaavian empiricism and also the traditional argument of chemists that their art cannot be reduced to mechanical causes. He was willing to accept geometrical methods of ordering and reasoning, the existence of matter as particles with varying shapes and motions, and the notion that these shapes and motions effect chemical properties. But specific chemical properties could only be determined empirically through chemical experimentation, not predicted via physical theory.[93]

Plummer and the Edinburgh Medical Faculty Founders

Soon after Crawford assumed the chair of Hebrew, he quit the teaching of chemistry altogether. One reason for this move was that by the early 1720s he faced competition from extramural lecturers. For his last advertised course in 1720, he partnered with Alexander Monro *primus*

(1697–1767), who taught anatomy for the Company of Surgeons, and Charles Alston (1683–1760), who was the King's Botanist.[94] By autumn of 1724, Crawford had competition from the foursome who would eventually become founding members of the university medical faculty – John Rutherford (1695–1779), Andrew St Clair (1697–1760), Andrew Plummer (1698–1756) and John Innes (1696–1733). By the next year, William Graeme (1701–1745) and George Martine (1702–1741) together gave courses in medicine and, probably, chemistry in competition with these four. All six had spent time in Leiden, and both groups had acquired facilities for their teaching and seem to have attracted significant numbers of students. The four won this competition when, along with Alexander Monro *primus*, they were appointed as professors in February 1726, officially marking the founding of the Edinburgh medical faculty. As Emerson argues, this victory was not solely due to the success of their medical teaching, but also to their political acceptability and to the practical and useful nature of their enterprise, both characteristics essential to securing the patronage of the Lord Provost, George Drummond (1688–1766) and the university's main patron, Archibald Campbell, Earl of Ilay (1682–1761).[95] Crawford simply could not, and did not wish to, compete with younger, better-equipped men for lecture fees. He continued to hold his chair and examine medical candidates until at least 1725, at which point he was removed from his chair.[96]

The four medical faculty founders managed their collaboration as a business partnership. In 1724 they jointly purchased a house next to the university's botanical garden, which they refitted into a chemical laboratory and lecture space. In addition to lecturing on medicine and chemistry, the partners ran a flourishing pharmaceutical business, in which they acquired large amounts of raw materials that they would process into medicaments for sale to local apothecaries.[97] This business made the partners wealthy and helped them to garner the support of their patron, the Earl of Ilay, who promoted commercial development in Scotland.[98] Each partner was supposed to teach chemistry in turn, but Andrew Plummer seems to have undertaken most of it. One set of student notes, which identifies Plummer as the instructor, survives from these early chemistry courses. These notes present 213 pharmaceutical recipes and two short lectures on essential oils and fermentation. Thus, the focus was overwhelmingly on practical applications. As Robert Anderson points out, Plummer and the partners were paid twice for their chemical work, since they collected lecture fees from students, who observed the preparation of the medicaments, which they later sold. Indeed, it was difficult to determine which enterprise, the selling of medicaments or teaching, was more important to them.[99]

Plummer's emphasis on the practical training and production of medicines seemed to run counter to the theoretical interests of Crawford and Boerhaave. On the surface, he appeared to be disinterested in Boerhaave's reform of chemistry, even though he was the only partner to have actually taken his medical degree at Leiden (1722), and all the partners advertised that their courses in medicine and chemistry were based on the 'system' of Herman Boerhaave.[100] In addition, Plummer garnered a reputation for practical instruction, much to the chagrin of some of his students, such as William Cullen (1710–1790), who argued that he needed to spend more time on theory.

Nevertheless, Plummer did have an interest in chemical theory, even if this interest is not immediately evident in surviving lecture notes. Two papers on the chemistry of solutions and salts, which he read to the Philosophical Society of Edinburgh in 1738 and 1739, reveal the theoretical dimension of his work.[101] His first paper, titled 'Remarks on chemical solutions and precipitations', examined the behaviour of solvents or *menstrua* and the displacement (precipitation) of chemical species in solution by other species, which have a greater 'affinity' for the *menstruum*. Plummer's general understanding of these reactions followed Boerhaave's approach by relying on the solidity, texture, shape and arrangement of the pores of solvent and solute particles to explain observed effects. In fact, Plummer, addressing a topic from Boerhaave's discussion of *menstrua*, deployed this model of chemical action to present a theoretical argument against the existence of the alkahest as a universal solvent. As he argued:

> there is a vast variety of bodies which differ from one another in density, solidity and texture, in the bulk, shape and composition of constituent particles, in the degree of force with which these cohere among themselves, and in the number, size, and figure of pores or interstices between the solid parts; it is scarce conceivable that any one liquor can be endued with powers corresponding to all the various circumstances and variety of bodies, so as indiscriminately to dissolve all.[102]

In Plummer's second paper, which described experiments that he made on acids and alkalis to generate various neutral salts, he utilised this same corpuscular matter theory, but also incorporated the notion of the 'attraction' and 'repulsion' of interacting particles as a further cause for the selective interaction of chemical species.[103] This model of attraction and repulsion was an important aspect of Boerhaave's later theory of *menstrua*, and he used this model to argue against a strict 'mechanical' interpretation of solubility (and displacement of species in solution) based on particle shape alone.[104] Plummer too argued that one cannot predict the effects of a *menstruum* on a specific body based on assumptions regarding particle shape, density, the knowledge of 'mechanical principles', or even on the previously observed properties of the *menstruum* and the solute body. The chemist must undertake experiments.[105]

Placing Plummer's two papers within the broader context of the chemical work done at the partners' laboratory suggests that his chemistry may have been misconstrued by later commentators. His first paper was very much organised according to a pedagogical method that followed the structure of Crawford's and Boerhaave's lectures on *menstrua*. Plummer structured his essay as a series of precepts, enumerating eight 'general canons' on the behaviour of *menstrua* followed by six 'remarks' on precipitations. He explained and supported his precepts with practical examples. Although he referred to some of these examples as universal experiences,[106] some of his examples seem to derive from demonstrations performed in a pedagogical setting. In one example he intended to show that 'bodies dissolved in their proper *menstrua*, many be precipitated thence, by several bodies of different qualities'. Plummer described an experiment in which he made a solution of 'pure silver' dissolved in 'good *aqua fortis*' that was then distributed into six or eight glasses. He then suggested that adding a variety of substances to these glasses – solutions of sea salt and crude sal ammoniac, spirit of sea salt, vitriol and sal ammoniac, fixed alkaline salt, and a piece of copper – will precipitate the silver.[107] As described, this experiment seems like a demonstration performed for a course, designed to reinforce the theoretical point about the unpredictability of the displacement of bodies in solution due to differing chemical affinities between solvents and solutes. It is likely that this paper was part of a lecture on the theory of *menstrua* that Plummer delivered to his students, but which simply has not survived in extant student notes.

Plummer organised his second paper very differently. He began the essay by describing a series of experiments which he had performed to separate the acid and alkaline parts of neutral salts by dissolving them in various acidic *menstrua*. Unlike his first essay, Plummer recounted experiments from the first-person perspective – 'I put two ounces and a half of this factitious nitre into a small retort [. . .]' – indicating that this was an experimental narrative rather than a pedagogical demonstration or statement of common experience.[108] At the end of his experiments, he summarised his conclusions in eight 'corollaries' followed by a 'general *scholium*'. While some of his conclusions follow from what he said in his first paper about *menstrua*, he also addressed some of the debates of his day. For example, he argued that acid salts retain their 'powers and produce the same effects', after they are separated from alkaline bases in neutral salts.[109] Some earlier chemists, notably Nicholas Lemery (1645–1715), argued that acids were blunted when reacting with alkalis and, as a result, some had their corrosive effects removed or reduced after separation.[110] Boerhaave, who discussed this issue at length in the *Elementa chemiae*, remained undecided, citing examples supporting both sides.[111] Ultimately, Plummer seemed to be moving away from the point/pore model of acid/alkali reactions favoured by early mechanical chemists, and replacing it with a model based on interparticulate forces of attraction and repulsion. At the end of the paper, he presented an affinity series for spirit of nitre (or *aqua fortis*): silver in a solution of *aqua fortis* is displaced by copper, which is in turn displaced by iron, etc., in what he called 'a series of bodies from silver to salt of tartar, whose attractive powers, with respect to the same saline liquor, are continually increasing'.[112] Although Boerhaave most likely introduced Plummer to the notion of interparticulate forces as the causes of solubility and chemical displacements, Plummer placed this model in a wider, Newtonian philosophical context by arguing that these forces account for many observed phenomena, such as Stephen Hales's work on 'the analysis of air' and the 'surprising experiments on electricity now so much in vogue'.[113]

Although the majority of Plummer's chemical work was practical, demonstrating recipes and refining simples into medicaments, he also engaged in teaching and research in chemical theory. Beginning with Boerhaave's ideas on *menstrua*, he began a programme of research into the solution chemistry of salts, which he attempted to explain using the idea of particle attraction and repulsion. Although he rarely used the word 'affinity', his work may have helped to establish the importance of affinity theory for later chemical instructors at Edinburgh, and also contributed to one of the main research agendas of the early eighteenth century.[114] Equally importantly, Plummer saw chemistry beyond the academic enterprise. This was evident in his key role in producing medicaments for the chemical business he founded with his medical faculty partners. Although his courses, at least later in his career, were much reviled by his successors, Plummer and his partners began the expansion of chemistry in Edinburgh away from solely medical applications. His successors to the chair of chemistry, William Cullen and Joseph Black (1728–1799), actively promoted industrial applications, and by the end of Black's teaching career, most of his audience was not composed of medical students.[115]

Conclusion

Herman Boerhaave provided a framework for academic chemistry at the University of Edinburgh. Both Crawford and Plummer adopted aspects of his method for ordering their courses and presenting the results of research. In addition, they adopted his philosophical perspective and used it as a starting point for their own work. Both rejected, explicitly or implicitly, the traditional chemical principles and speculative approach of 'mechanical' chemistry. Instead, they advocated an empirical approach in which chemical phenomena, as latent properties of matter, must be investigated through systematic experimentation. The results of these experiments were ordered and interpreted using a version of Boerhaave's instruments theory. This framework, established in Edinburgh by Crawford and Plummer, continued to shape the chemistry teaching and research of their successors. Cullen and Black, for example, both extended Boerhaave's thermometry experiments in new and fruitful directions, but began their work by combing the pages of the *Elementa Chemiae*.[116]

This did not mean, however, that the development of chemistry in Edinburgh was simply an extension of Boerhaave's programme. Rather, it was a starting point. Crawford and Plummer pursued their chemistry in response to their local circumstances. For Crawford, this meant a strong engagement with the British and Scottish Newtonians, while Plummer worked at making medicaments to establish and keep afloat the fledgling Edinburgh medical faculty. Neither, it would seem, had the strong interest in experimental anatomy or physiological theory which was a central part of medicine at Leiden and in Boerhaave's own work. This move away from medicine and towards both theoretical development and practical, industrial application shaped the pursuit of chemistry in Edinburgh through the eighteenth century. William Cullen called this kind of chemistry 'philosophical chemistry', which he saw as a form of natural philosophy, independent from medicine.[117] Itinerant lecturer Peter Shaw (1694–1763) best defined this notion as chemistry pursued 'by means of appropriate experiments, scientifically explained, lead[ing] to the discovery of *Physical Axioms*, and *Rules of Practice*, for producing useful effects' in order to 'improve the state of natural knowledge, and the arts thereon depending'.[118] Boerhaave too was part of this philosophical tradition, and he argued for the relevance of chemistry in natural philosophy and its utility in many practical arts.[119] Yet he undertook few practical projects outside of medicine; only in Scotland was Shaw's philosophical chemistry fully realised.

Notes and References

1　On the founding of the Edinburgh medical school and Boerhaave's influence, see Guthrie, D., 'The Influence of the Leyden School upon Scottish Medicine', *Medical History* 3 (1959), pp. 108–22; Anderson, R.G.W. and A.D.C. Simpson (eds), *The Early Years of the Edinburgh Medical School* (Edinburgh: Royal Scottish Museum, 1976); Underwood, E.A., *Boerhaave's Men at Leyden and After* (Edinburgh: Edinburgh University Press, 1977), pp. 88–125; Cunningham, A., '"Medicine to Calm the Mind": Boerhaave's Medical System and Why It Was Adopted in Edinburgh', in A. Cunningham and R. French (eds), *The Medical Enlightenment of the Eighteenth Century* (Cambridge: Cambridge University Press, 1990), pp. 40–66. For an insightful critique of the traditional historiography of the founding, see Emerson, R., 'The Founding of the Edinburgh Medical School', *Journal of the History of Medicine and Allied Sciences*, 59 (2004), pp. 183–218.

2　Clow, A., 'Herman Boerhaave and Scottish Chemistry' in A. Kent (ed.), *An Eighteenth-Century Lectureship in Chemistry* (Glasgow: University of Glasgow Press, 1950), pp. 41–8. For a more nuanced view, see Anderson, R.G.W., 'Boerhaave to Black: The Evolution of Chemistry Teaching', *Ambix* 53 (2006), pp. 237–54.

3　Quoted in Cunningham, 'Medicine to Calm the Mind', p. 41.

4　For Boerhaave as empiricist, see Cook, H.J., 'Boerhaave and the Flight from Reason in Medicine', *Bulletin of the History of Medicine* 74 (2000), pp. 221–40; as mechanist, see King, L.S., *The Medical World of the Eighteenth Century* (Chicago, IL: Chicago University Press, 1958), Chs. 1–2; as Calvinist, see Knoeff, R., *Herman Boerhaave (1668–1738): Calvinist*

Chemist and Physician (Amsterdam: Koninklijke Nederlandse Akademie van Wetenschappen, 2002); as Newtonian, see Thackray, A., *Atoms and Powers: An Essay on Newtonian Matter-Theory and the Development of Chemistry* (Cambridge: Cambridge University Press, 1970), pp. 106–13.

5 On Leiden's medical school and institutional resources, see Lindeboom, G.A., 'Medical Education in the Netherlands, 1575–1760' in C.D. O'Malley (ed.), *The History of Medical Education* (Berkeley: University of California Press, 1970), pp. 201–16; Lunsingh Scheurleer, Th.H. and G.H.M. Posthumus Meyjes (eds), *Leiden University in the Seventeenth Century: An Exchange of Learning* (Leiden: Universitaire Pers Leiden/Brill, 1975); Heninger, J., 'Some Activities of Herman Boerhaave, Professor of Botany and Director of the Botanic Garden at Leiden', *Janus* 58 (1971), pp. 1–78; Beukers, H., 'Clinical Teaching at Leiden from Its Beginning until the End of the Eighteenth Century', *Clio Medica* 21 (1987–8), pp. 139–52.

6 See Cook, H.J., *Trials of an Ordinary Doctor: Johannes Groeneveldt in Seventeenth-Century London* (Baltimore, MD: Johns Hopkins University Press, 1994), p. 63.

7 On Harvey's ideas in Leiden, see French, R., 'Harvey in Holland: Circulation Among the Calvinists', in R. French and A. Wear (eds), *The Medical Revolution of the Seventeenth Century* (Cambridge: Cambridge University Press, 1989), pp. 46–86. For Descartes' theories in Leiden, see: Verbeek, T., *Descartes and the Dutch: Early Reaction to Cartesian Philosophy, 1637–1650* (Carbondale and Edwardsville, IL: Southern Illinois University Press, 1992); McGahagan, T., 'Cartesianism in the Netherlands, 1639–1676: The New Science and the Calvinist Counter-Reformation', PhD Thesis, University of Pennsylvania, 1976; Luyyendijk-Elshout, A.M., '*Oeconomia Anamalis*, Pores and Particles: The Rise and Fall of the Medical Philosophical School of Theodoor Craanen (1621–1690)' in Th.H. Lunsingh Scheurleer and G.H.M. Posthumus Meyjes (eds), *Leiden University in the Seventeenth Century*, pp. 295–307.

8 On religious toleration in the Dutch Republic and university, see Berkvens-Stevelinck, C., J. Israel and G.H.M. Posthumus Meyjes (eds), *The Emergence of Toleration in the Dutch Republic* (Leiden: Brill, 1997); Wolter, J.J., 'Introduction', in Th.H. Lunsingh Scheurleer and G.H.M. Posthumus Meyjes (eds), *Leiden University in the Seventeenth Century*, pp. 1–19. Regarding the Greek student, Boerhaave received a letter of introduction for this student from the Patriarch of Constantinople; see Lindeboom, G.A. (ed.), *Boerhaave's Correspondence*, Part II (Leiden: Brill, 1964), pp. 6–11.

9 On Sylvius's medical theories, see Baumann, E.D., *François Dele Boë Sylvius* (Leiden: Brill, 1949); King, L., *Road to Medical Enlightenment, 1650–1695* (New York: Elsevier, 1970), pp. 93–113; Underwood, E.A., 'Franciscus Sylvius and his Iatrochemical School', *Endeavour* 31 (1972), pp. 73–6; Beukers, H., 'Acid Spirits and Alkaline Salts: The Iatrochemistry of Franciscus dele Boë, Sylvius', *Sartoniana* 12 (1999), pp. 39–58.

10 See Boerhaave, H., *Oratio de suo errores expurgante* (Leiden, 1718). An English translation of this oration, 'Discourse on Chemistry Purging Itself of Its Own Errors', is found in Kegel-Brinksgreve, E. and A.M. Luyendijk-Elshout (eds), *Boerhaave's Orations* (Leiden: Brill, 1983), pp. 180–213.

11 See Ragland, E.R., 'Experimenting with Chymical Bodies: Reinier de Graaf's Investigations of the Pancreas', *Early Science and Medicine* 13 (2008), pp. 615–64; Ragland, E.R., 'Chemistry and Taste in the Seventeenth Century: Franciscus Dele Boë Sylvius as a Chymical Physician Between Galenism and Cartesiansim', *Ambix* 59 (2012), pp. 1–21. For Boerhaave's use of chemical analysis in anatomy and physiology, see Powers, J.C., *Inventing Chemistry: Herman Boerhaave and the Reform of the Chemical Arts* (Chicago, IL: University of Chicago Press, 2012), pp. 106–14.

12 For Schacht's course, see British Library, Sloane Collection, MS 1287; Powers, *Inventing Chemistry*, pp. 48–9.

13 On De Maets, see Van Spronsen, J.W., 'The Beginning of Chemistry', in Th.H. Lunsingh Scheurleer and G.H.M. Posthumus Meyjes (eds), *Leiden University in the Seventeenth Century*, pp. 329–43 (on pp. 336–8); Lindeboom, G.A., 'Maets (Dematius), Carel', *Dutch Medical Biography* (Amsterdam: Rodopi, 1984), pp. 260–1.

14 Morley, C.L., *Collectanea Chymica Lydensia, id est Matensiana, Margraviana et le Mortiana; Silicet trium in Academia Lugduno-Batava facultatis chimiae* (Lugduni Batavorum, Henricum Drummond, 1684). On the three competing lecturers and De Maets's status, see Powers, *Inventing Chemistry* (n.11), pp. 50–5; Van Spronsen, 'Beginning of Chemistry', pp. 338–41.

15 'Advies van de Medische faculteit over privaat-colleges van niet-professoren,' 22 March 1690, in P.C. Molhuysen (ed.), *Bronnen tot de Geschiedenis der Leidsche Universiteit*, vol. 6 ('sGravenhage: Martinus Nijhoff, 1921), p. 23; Powers, *Inventing Chemistry* pp. 55–6; Lindeboom, G.A., *Herman Boerhaave: The Man and His Work* (London: Meuthen, 1968), pp. 109–10.

16 For Boerhaave's Education in medicine, see Lindeboom, *Herman Boerhaave*, pp. 25–7; Powers, *Inventing Chemistry*, pp. 23–5; Cunningham, 'Medicine to Calm the Mind', pp. 46–51.

17 See Burton, W., *An Account of the Life and Writings of Herman Boerhaave*, 2nd edn (London: Henry Lintot, 1746), p. 16; Powers, *Inventing Chemistry*, pp. 57–8. On Stam, see Lindeboom, G.A., 'David en Nicholaas Stam, apothekers te Leiden', *Pharmaceutisch Weekblad* 108 (1973), pp. 153–60.

18 Penning, C.P.J., 'De Promotie van Boerhaave te Harderwijk', *Nederlandsch Tijdschrift voor Geneeskunde* 82 (1938), pp. 4895–9; Lindeboom, *Herman Boerhaave*, pp. 38–42.

19 On the 'canal boat incident', see Knoeff, *Herman Boerhaave*, pp. 30–46; Lindeboom, *Herman Boerhaave*, pp. 45–7.

20 Some early experimental notes from this period exist, mainly on transmutational alchemy – see Fundamental Library, Military Medicine Academy, St Petersburg, Russia (Fundamental'naya Biblioteka, Voenno-Meditsinskoi

Akademii, hereafter VMA), Fund XIII, MS 1, ff. 85r–86v.

21 Resolutiones Curatores, 4 January 1702, in Molhuysen (ed.), *Bronnen*, vol. 4, p. 190. On Boerhaave's appointments, see Powers, *Inventing Chemistry*, pp. 27–8.

22 See Van Poelgeest, L., 'The Stadholder-King William III and the University of Leiden', in P. Hoftijzer and C.C. Barfoot (eds), *Fabrics and Fabrications: The Myth and Making of William and Mary* (Amsterdam and Atlanta: Rodopi, 1990), pp. 97–134; Lindeboom, *Herman Boerhaave*, pp. 51–2.

23 Powers, *Inventing Chemistry*, pp. 28–9. Quote from Res. Cur. 12 April & 8 May 1703, in Molhuysen (ed.), *Bronnen*, vol. 4, p. 208. On the minor controversy over Boerhaave's appointment, because he had never taught botany, see Ultee, M., 'The Politics of Professorial Appointment at Leiden, 1709', *History of Universities* 9 (1990), pp. 167–94.

24 Powers, *Inventing Chemistry*, pp. 65–8.

25 Morley, *Collectanea Chymica*.

26 Lemery, N., *Cours de Chymie, contenant la maniere de faire les Operations qui sont en usage dans la Medicine, par une Methode facile*, 10th edn (Paris: Jean-Baptiste Delespina, 1713).

27 Boerhaave, H., *Elementa chemiae*, 2 vols (Lugdunum Batavorum, Isaacum Severinum, 1732). For this paper, I use the reliable translation: Boerhaave, H., *Elements of Chemistry*, trans. Timothy Dallowe, 2 vols (London: J. and J. Pemberton, 1735).

28 For De Maets's principles, see 'Collegium chemicum', VMA, MS 131, f. 56r; for Lemery, see Lemery, *Cours de chymie*, pp. 2–31.

29 On the critique of chemical principles, see Debus, A.G., 'Fire Analysis and the Elements in the Sixteenth and Seventeenth Centuries', *Annals of Science* 23 (1967), pp. 127–47; Principe, L.M., *The Aspiring Adept: Robert Boyle and His Alchemical Quest* (Princeton, NJ: Princeton University Press, 1998), pp. 27–62. Boyle's influential critique is found in Boyle, R., *The Sceptical Chymist* (London, 1661); Boyle, R., *The Producibleness of Chymical Principles* (Oxford, 1680). On the impact of Boyle's arguments, see Clericuzio, A., 'Carnaedes and the Chemists: A Study of *The Sceptical Chymist* and its Impact on Seventeenth-century Chemistry', in M. Hunter (ed.), *Robert Boyle Reconsidered* (Cambridge: Cambridge University Press, 1994), pp. 79–88; Clericuzio, A., '"Sooty Empiricks" and Natural Philosophers: The Status of Chemistry in the Eighteenth Century', *Science in Context* 23 (2010), pp. 329–50.

30 Boerhaave, *Elements*, vol. I, pp. 78–500.

31 Bohn composed these dissertations for his medical students to debate in public disputation exercises; the student *defendens* is named at the start of each dissertation. See Bohn, J., *Dissertationes Chymico-Physicae* (Lipsae: Joh. Fredericus Gleditschius, Literis Christianus Gözl, 1685). Boerhaave took extensive notes from Bohn's dissertations; see VMA, MS 3, ff. 94r–117r. On the origins and development of the instrument theory, see Powers, *Inventing Chemistry*, pp. 74–7.

32 VMA, MS 3, ff. 28r–v. On the French programme, see Principe, L.M., 'Wilhelm Homberg: Chymical Corpuscularianism and Chrysopoeia in the Early Eighteenth Century', in C. Lüthy, J.E. Murdoch and W.R. Newman (eds), *Late Medieval and Early Modern Corpuscular Matter Theories* (Leiden: Brill), pp. 535–56 (on pp. 539–46); Principe, L.M., 'Wilhelm Homberg et la chimie de la lumière', *Methodos* [online] 8 (2008), http://methodos.revues.org/1223 (last accessed 25 March 2015).

33 VMA, MS 3, ff. 28v.

34 See Powers, *Inventing Chemistry*, ch.3.

35 Boerhaave, H., *Oratio de Commendando Studio Hippocratio* (Leiden, 1701). See also Kegel-Brinksgreve, E. and A.M. Luyendijk-Elshout (eds), *Boerhaave's Orations* (Leiden: Brill, 1983), pp. 54–84.

36 The best description of this model is found in Boerhaave's instruments course; see VMA, MS 7, ff. 1v–2r. See also Powers, *Inventing Chemistry*, pp. 118–21; Knoeff, R., 'Practising Chemistry "After the Hippocratical Manner": Hippocrates and the Importance of Chemistry in Boerhaave's Medicine', in L. Principe (ed.), *New Narratives in Eighteenth-Century Chemistry* (Dordrecht: Springer, 2007), pp. 63–76. On Baconian fact-collecting, see the classic: Webster, C., *The Great Instauration: Science, Medicine, and Reform* (London: Duckworth, 1975).

37 Boerhaave, *Elements*, vol. I, pp. 214–22. For another view of Baconian 'histories' in Boerhaave, see Klein, U., 'Experimental History and Herman Boerhaave's Chemistry of Plants', *Studies in History and Philosophy of Science, Part C: Studies in History and Philosophy of Biological and Biomedical Sciences* 34 (2003), pp. 533–67.

38 VMA, MS 3, f. 38v. Note this 'mechanical' view of chemical interactions was common; see for example, Lemery, *Cours de chymie*, pp. 22–5; Homberg, W., 'Essais de Chimie, Article Premier: Des Principes de la Chimie in general', *Memoires de l'Académie Royale des Sciences, année 1702* (Amsterdam, 1737), pp. 44–8.

39 VMA, MS 3, ff. 38r–40r.

40 VMA, MS 3, ff. 38r.

41 The 'instruments course' lasted until 1728; see VMA, MS 7; Powers, *Inventing Chemistry*, ch.5.

42 On the 'method of the geometers', see Powers, *Inventing Chemistry*, pp. 102–3, 121–9. For an example, see Boerhaave's demonstrations on the expansion of bodies when heated: VMA, MS 7, ff. 3r–4v; Boerhaave, *Elements*, vol. I, pp. 86–107. On Newton's influence, see Dear, P., *Discipline and Experience: The Mathematical Way in the Scientific Revolution* (Chicago, IL: University of Chicago Press, 1995), pp. 210–27; Garrison, J.W., 'Newton and the Relation of Mathematics to Natural Philosophy', *Journal of the History of Ideas* 48 (1987), pp. 609–27.

43 Boerhaave, H., 'De Mercurio Experimenta', *Philosophical Transactions of the Royal Society of London* 38 (1733), pp. 145–76, 343–59 and 368–76. See

also Powers, J.C., 'Scrutinizing the Alchemists: Herman Boerhaave and the Testing of Chymistry', in L.M. Principe (ed.), *Chymists and Chymistry: Studies in the History of Alchemy and Early Modern Chemistry* (Sagamore Beach, MA: Chemical Heritage Foundation & Science History Publications), pp. 227–38; Powers, *Inventing Chemistry*, pp. 179–90.

44 On the notion of 'philosophical instruments' see Warner, D.G., 'What Is a Scientific Instrument, When Did It Become One, and Why?', *British Journal for the History of Science* 23 (1990), pp. 83–93.

45 Boerhaave, *Elements*, vol. I, pp. 222, 292.

46 Boerhaave, *Elements*, vol. 2. Method is discussed on pp. 1–2.

47 Lindeboom, G.A., *Bibliographia Boerhaaviana* (Leiden: Brill, 1959), pp. 81–6.

48 Lindeboom, *Herman Boerhaave*, p. 356.

49 Lindeboom, *Herman Boerhaave*, pp. 360–74.

50 Helen Dingwald estimated that 150 surgical apprentices worked in Edinburgh c. 1700; see Dingwald, H., *Physicians, Surgeons and Apothecaries: Medicine in Seventeenth Century Edinburgh* (East Linton: Tuckwell Press, 1995), pp. 25, 70.

51 See Emerson, 'Founding', pp. 188.

52 Emerson, 'Founding', p. 189; Anderson, R.G.W., *The Playfair Collection and the Teaching of Chemistry at the University of Edinburgh 1713–1858* (Edinburgh: Royal Scottish Museum, 1978,), p. 4.

53 Anderson, *Playfair Collection*, pp. 3–4.

54 Underwood, E.A., *Boerhaave's Men at Leyden and After* (Edinburgh: Edinburgh University Press, 1977), pp. 94–5.

55 Cunningham, A., 'Robert Sibbald and Medical Education, Edinburgh, 1706', *Clio Medica* 13 (1979), pp. 135–61. On the circulation of Pitcairne's course notes, see Underwood, *Boerhaave's Men*, pp. 203–4, n. 113. This discusses the medical course notes found in Wellcome MS 2451.

56 Emerson, 'Founding'.

57 Emerson, 'Founding', pp. 189–93.

58 Underwood, *Boerhaave's Men*, pp. 89–91; Cunningham, 'Robert Sibbald'. On Pitcairne's Leiden tenure, see Lindeboom, G.A., 'Pitcairne's Leyden Interlude Described from the Documents', *Annals of Science* 19 (1963), pp. 273–84.

59 Underwood, *Boerhaave's Men*, pp. 95–7, 102–6; St Clair, R. E.W., *The Doctors Monro: A Medical Saga* (London: Wellcome Historical Medical Library, 1964). Note that Roger Emerson has recently revised the original account of Monro's influence; see Emerson, 'Founding'.

60 Pitcairne, for example, replaced the exam on 'several material questions' with one on the 'Institutes of Medicine', the standard theory course in Leiden. See Underwood, *Boerhaave's Men*, p. 91. For the Leiden exams, cf. Lindeboom, 'Medical Education in the Netherlands', p. 203.

61 On Crawford's early career and education, see Underwood, *Boerhaave's Men*, pp. 99–100; Anderson, *Playfair Collection*, p. 4.

62 Emerson, 'Founding', p. 202.

63 Anderson, *Playfair Collection*, p. 4.

64 Emerson, 'Founding', pp. 202–3.

65 Underwood, *Boerhaave's Men*, p. 100.

66 Gibson, J., 'An extraordinary large Gall-Bladder and hydropick Cystis' in *Medical Essays and Observations*, vol. II, 3rd edn (Edinburgh: W. & T. Ruddimans, 1747), pp. 299–304 (on pp. 301–2). On Crawford's examining medical students, see Underwood, *Boerhaave's Men*, pp. 101–2.

67 Underwood, *Boerhaave's Men*, pp. 101, 203–4, n. 113; Anderson, *Playfair Collection*, pp. 4–5.

68 Anderson, *Playfair Collection*, p. 4.

69 Underwood, *Boerhaave's Men*, p. 106; Anderson, *Playfair Collection*, p. 5.

70 These notes are held at the Wellcome Library, London, MS 2451 (hereafter Wellcome 2451). The notes, separately paginated from the rest of the manuscript, are titled 'Tractatus Chymici (a Doctore Crawford Dictati)', and give an end date of June 1713. Anderson suggests that this date is erroneous, since Crawford was not appointed until December; see Anderson, *Playfair Collection*, p. 15, n. 29. I compare these notes to Boerhaave's earliest courses, VMA, MS 3.

71 Boerhaave taught chemistry as a private course, meaning he collected lecture fees from auditors, who may not have matriculated, during every winter term from 1705–11. Cf. VMA, MS 3, 1r, where he recorded the start date of each chemistry course.

72 Cf. Wellcome 2451, p. 14; VMA, MS 3, 18r. Boerhaave cribbed his reported values for the specific weight of metals from a report published in the *Philosophical Transactions of the Royal Society of London*: Anon., 'A Further List of the Specific Gravities of Bodys, being in proportion as the following Numbers', *Philosophical Transactions of the Royal Society of London* 15, no. 169 (Jan. 1685), pp. 927–9.

73 Wellcome 2451, p. 1.

74 Wellcome 2451, p. 2. This was also a theme of Boerhaave's course, especially in his history of chemistry, and later, his oration, *De Chemia suos errors expurgante*, which he presented in 1718 on his appointment to the chair of chemistry at Leiden; see Kegel-Brinksgreve, E. and A.M. Luyendijk-Elshout (eds), *Boerhaave's Orations*, pp. 180–213.

75 Wellcome 2451, pp. 4–5.

76 Wellcome 2451, pp. 43–111; discussion of method on pp. 43–4. For Boerhaave's statement of the pedagogical expediency of this method, see Boerhaave, *Elements*, Vol II, pp. 1–2.

77 Wellcome 2451, pp. 13–21.

78 Wellcome 2451, p. 21.

79 Wellcome 2451, p. 23.

80 Wellcome 2451, p. 4. For Crawford's discussion of atomism, see p. 22.

81 Wellcome 2451, pp. 24–8. Note Ursula Klein calls these kinds of principles

'proximate principles'; they are used in a practical way, as the products of analysis, with the notion of their elemental simplicity removed. See Klein, U. and W. Lefèvre, *Materials in Eighteenth-Century Science: A Historical Ontology* (Cambridge, MA: MIT Press, 2007), pp. 221–47.

82 Wellcome 2451, p. 24.

83 Cf. Wellcome 2451, pp. 37–41; VMA, MS 3, ff.38r–43r. On the alkahest, see Porta, P.A., "*Summus atque felicissimus salium*": The Medical Relevance of the Liquor Alkahest', *Bulletin of the History of Medicine* 76 (2002), pp. 1–29; Joly, B., 'L'alkahest, dissolvant universel ou quand la théorie rend pensible un pratique impossible', *Revue d'Histoire des Sciences* 49 (1996), pp. 305–44.

84 Cf. Wellcome 2451, pp. 37–8; VMA, MS 3, ff.38r–v. Note that Boerhaave later adds a fourth factor, a 'reciprocal virtue' or 'attraction' between the *menstruum* and a body – what later chemists will call chemical 'affinity'; see Boerhaave, *Elements*, vol. I, pp. 390–5.

85 Wellcome 2451, p. 37.

86 Cf. Wellcome 2451, pp. 29–10, 31; VMA, MS 3, ff.26r–v.

87 Newton, I., *Opticks, or A Treatise of the Reflections, Refractions, Inflections and Colours of Light* (London: Sam Smith and Benjamin Walford, 1704). The theory of light as a particle is found in the 'queries'; see Home, R.W., 'Newton's Subtle Matter: the *Opticks* Queries and the Mechanical Philosophy', in J.V. Field and F.A.J.L. James (eds), *Renaissance and Revolution: Humanists, Scholars, Craftsmen, and Natural Philosophers in Early Modern Europe* (Cambridge: Cambridge University Press, 1993), pp. 193–202.

88 Cf. Wellcome 2451, p. 30; VMA, MS 3, ff. 28 r–v. See also Boerhaave, *Elements*, vol. I, pp. 125–52.

89 Wellcome 2451, p. 30. This theory derives from Query 10 of Newton's *Opticks* (1704 edn); see Newton, *Opticks*, Bk. 2, p. 134.

90 See Guerrini, A., 'Archibald Pitciarne and Newtonian Medicine', *Medical History* 31 (1987), pp. 70–83; Guerrini, A., 'James Keill, George Cheyne, and Newtonian Physiology, 1690–1740', *Journal of the History of Biology* 18 (1985), pp. 247–66; Brown, T.M., 'Medicine in the Shadow of the *Principia*', *Journal of the History of Ideas* 48 (1987), pp. 629–48.

91 Freind, J., *Chymical Lectures, In which almost all the Operations of Chymistry are Reduced to their True Principles, and the Laws of Nature* (London: Phillip Gwillim for Jonah Bowyer, 1712), pp. 1, 7–10. On Freind and chemistry, see Rowlinson, J.S., 'John Freind: Physician, Chemists, Jacobite, and Friend of Voltaire's', *Notes and Records of the Royal Society of London* 61 (2007), pp. 109–27; Guerrini, A., 'Chemistry Teaching at Oxford and Cambridge, circa 1700' in P.M. Rattansi and A. Clericuzio (eds), *Alchemy and Chemistry in the Sixteenth and Seventeenth Centuries* (Dordrecht: Kluwer, 1994), pp. 183–99; Thackray, *Atoms and Powers*, pp. 70–3. For Newton's *De natura acidorum*, see Turnbull, H.W. et al. (eds), *The Correspondence of Isaac Newton*, vol. III (Cambridge: Cambridge University Press, 1961), pp. 202–14.

92 Wellcome 2451, p. 3.

93 Wellcome 2451, pp. 3–4. On the demarcation of chemistry from physics in the eighteenth century, see Boantza, V.D., *Method and Matter in the Long Chemical Revolution: Laws of Another Order* (Farnham: Ashgate, 2013); Meinel, C., 'Theory or Practice?: The Eighteenth-Century Debate on the Scientific Status of Chemistry', *Ambix* 30 (1983), pp. 121–32.

94 Anderson, *Playfair Collection*, p. 5; Underwood, *Boerhaave's Men*, p. 106.

95 Emerson, 'Founding', pp. 205–16.

96 Emerson, 'Founding', p. 205; Anderson, *Playfair Collection*, p. 5.

97 Anderson, *Playfair Collection*, pp. 8–9. Anderson states, for example, that in 1738, Rutherford ordered 56 pounds of cinnamon (p. 8).

98 Emerson, 'Founding', pp. 214–15. On the Early of Ilay's commercial and scientific interests, see Emerson, R.L., 'The Scientific Interests of Archibald Campbell, 3rd Duke of Argyll (1682–1761)', *Annals of Science* 59 (2002), pp. 21–56.

99 See Anderson, 'Boerhaave to Black', pp. 247–8; Anderson, *Playfair Collection*, p. 9. Notes from Plummer's course are held at the Royal College of Physicians of Edinburgh, MS M8/17–20. Anderson suggests that a second set of notes from Plummer's course may be held at Glasgow University Library Special Collections, MS Hunter H487.

100 For an extended discussion of this, see Cunningham, 'Medicine to Calm the Mind'.

101 Plummer, A., 'Remarks on Chemical Solutions and Precipitations', in *Essays and Observations, Physical and Literary*, vol. 1 (Edinburgh: Philosophical Society of Edinburgh, 1754), pp. 284–314; Plummer, A., 'Experiments on Neutral Salts, compounded of different acid liquors, and Alkaline Salts, fixed and volatile', in *Essays and Observations*, pp. 315–40.

102 Plummer, 'Remarks', pp. 286–7.

103 Cf. Plummer, 'Experiments', pp. 331–2: 'the fluid acid salt of the oil of vitriol attracts one part of each real particle of nitre, *viz.* the fixt and alkaline basis, while this same vitriolic acid seems to repel another part of nitre, that is the acid and volatile part . . .'

104 Cf. Boerhaave, *Elements*, vol. I, pp. 391–2 and 397–401. See also Duncan, A., *Laws and Order in Eighteenth-Century Chemistry* (Oxford: Clarendon Press, 1996), pp. 56–9.

105 Plummer, 'Remarks', pp. 287–90. For an argument against 'mechanical principles', see pp. 305–6.

106 See, for example, the following example, presented as a common fact: 'Iron and copper, the hardest of metals, which require the greatest force to extend them, and the strongest fire to bring them into fusion, will be corroded and dissolved by liquors most harmless to the human body; as vinegar, juice of lemons, a solution of tartar, rhenish wine, nay moist air . . .' Plummer, 'Remarks', p. 288.

107 Plummer, 'Remarks', p. 309.

108 Plummer, 'Experiments', p. 316.

109 Plummer, 'Experiments', p. 328. The experiments to validate this claim are found on pp. 323–4.

110 See Lemery, *Cours de chymie*, pp. 24–5. For Lemery, the blunting of acids, which he asserted were composed of particles with sharp points, occurred when the acid point broke off during interaction with the pores of alkalis. On Lemery's corpuscular interpretation of acid/alkali reactions, see Bougard, M., *La Chimie du Nicolas Lemery* (Turnhout: Brepols, 1999).

111 Boerhaave, *Elements*, vol. I, pp. 411–12, 470–2. Boerhaave's discussion of this issue was undoubtedly the root of Plummer's interest in it.

112 Plummer, 'Experiments', pp. 333–7, quote on p. 337. Note that Plummer does not use the word 'affinity' here, but merely 'attraction'; however, these concepts were coming to mean the same thing. See Duncan, *Laws and Order*, pp. 96–104.

113 Plummer, 'Experiments', pp. 339–40. On 'Newtonian' attraction and repulsion in Hales's work, see Thackray, *Atoms and Powers*, pp. 114–18; Duncan, *Laws and Order*, pp. 76–7; Guerlac, H., 'The Continental Reputation of Stephen Hales', *Archives Internationales d'Histoire des Sciences* 4 (1951), pp. 393–404. On these forces in electrical experimentation, see Heilbron, J., *Elements of Early Modern Physics* (Berkeley: University of California Press, 1982), ch. III.

114 On the importance of salt chemistry and affinity theory, see Holmes, F.L., *Eighteenth Century Chemistry as an Investigative Enterprise* (Berkeley, CA: Office for the History of Science and Technology, University of California, 1982), ch. 2; Kim, M.G., *Affinity, That Noble Dream: A Genealogy of the Chemical Revolution* (Cambridge, MA: MIT Press, 2003). For the importance of affinity theory in chemistry teaching at Edinburgh, see Taylor, G., 'Marking Out a Disciplinary Common Ground: The Role of Chemical Pedagogy in Establishing the Doctrine of Affinity at the Heart of British Chemistry', *Annals of Science* 65 (2008), pp. 465–86.

115 Golinski, J., *Science as Public Culture: Chemistry and Enlightenment in Britain, 1760–1820* (Cambridge: Cambridge University Press, 1992), ch. 2; Donovan, A.L., *Philosophical Chemistry in the Scottish Enlightenment: The Doctrines and Discoveries of William Cullen and Joseph Black* (Edinburgh: Edinburgh University Press, 1975). On the audience for Black's courses, see Anderson, 'Boerhaave to Black', pp. 250–1.

116 Guerlac, H., 'Joseph Black's Work on Heat' in A.D.C. Simpson (ed.), *Joseph Black: 1728–1799: A Commemorative Symposium* (Edinburgh: Royal Scottish Museum, 1982), pp. 13–22; Cullen, W., 'Of the Cold Produced by Evaporating Fluids, and Some Other Means of Producing Cold' in *Essays and Observations, Philosophical and Literary*, vol. 2 (Edinburgh, 1770), pp. 159–71. For Boerhaave's work on thermometry, see Powers, J.C., 'Measuring Fire: Herman Boerhaave and the Introduction of Thermometry into Chemistry', *Osiris* 29 (2014), pp. 158–77.

117 Donovan, *Philosophical Chemistry*; Christie, J.R.R., 'William Cullen and the Practice of Chemistry' in A. Doig, J.P.S. Ferguson, I.A. Milne and R. Passmore (eds), *William Cullen and the Eighteenth-Century Medical World* (Edinburgh: Edinburgh University Press, 1993), pp. 98–109; Taylor, G., 'Unification Achieved: William Cullen's Theory of Heat and Phlogiston as an Example of His Philosophical Chemistry', *British Journal for the History of Science* 39 (2006), pp. 477–501.

118 Shaw, P., *Chemical Lectures, Publickly Read at London in the Years 1731 and 1732, and since at Scarborough, in 1733* (London: J. Shuckburgh and Tho. Osborne, 1734), p. 1.

119 See Boerhaave, *Elements*, vol. I, pp. 52–72; Powers, *Inventing Chemistry*, pp. 4–8 and 141–69.

FOUR

Plummer to Cullen:
Novelty in William Cullen's Chemical Pedagogy

GEORGETTE TAYLOR

William Cullen's predecessor, Andrew Plummer, had a career of almost 30 years lecturing in chemistry at Edinburgh. Considering Cullen's teaching provides an opportunity to re-examine and compare that of his predecessor, as well to try and throw some new light on our historical assessment of their respective roles in the chemical pedagogy of the eighteenth century. The task has prompted some valuable reflections on what we are legitimately able to draw from different sources, and the way in which factors beyond the sources themselves can affect our interpretation of the evidence which we have.

Andrew Plummer has often been seen by historians up to the present as simply being William Cullen's predecessor. The impression invariably left is that his chemistry teaching was not as good as Cullen's. The questions could be asked, what was it about Cullen's chemistry teaching that so cast Plummer's into the shade, and is the attitude adopted to Plummer by historians justified on the basis of the evidence? There is one quote that tends to be ubiquitous in any discussion of Plummer's career, and indeed that has served for a host of historians to characterise Plummer's pedagogical endeavours. This is, arguably, the main source for our general impression that Plummer was an inferior lecturer to Cullen. In a letter to William Cullen of 1755, Joseph Black, Cullen's former student, and erstwhile contender to become Plummer's successor, wrote: 'you need not be anxious, provided that your course be better than Plummer's, which it is impossible for it not to be'.[1]

The character of Plummer's teaching appears to have been set for posterity by this comment. Or at least since 1832, when this letter was published for the first time in John Thomson's *Life and Work of William Cullen*.[2] Black's comment was, of course, his personal expressed opinion, and there are reasons for caution in accepting his words at face value. Black's thoughts were being expressed in private correspondence to

Cullen, who was at that time attempting to wrestle Plummer's chair in chemistry at Edinburgh away from him, Plummer having suffered a stroke.[3] Black was also in the somewhat difficult situation of being preferred by both Plummer's family and the medical faculty to take over the chemistry teaching during Plummer's indisposition. He was thus navigating a tortuous route between friendship, respect and honesty, telling Cullen what he would presumably prefer to hear, and not anticipating the later publication of his private comments. The view might be taken that Black was perhaps being harsher on Plummer than he deserved in order to gratify Cullen, safe in the knowledge that his comments would not come to Plummer's attention. Reading Black's comment in another way, the lines might be seen as not quite so flattering to Cullen, with whose teaching at Glasgow Black was closely familiar. There really should have been no need at all to include the term 'provided that'. Similarly, the 'which it is impossible for it not to be' is arguably something of a backhanded compliment. Black and Plummer must have had a reasonable relationship in reality, although it is not surprising if this was not reflected with absolute clarity in his letters to Cullen. We can probably conclude that the cautious historian might not choose to take Black's comment, given to whom it was made, and when it was made, as necessarily a careful judgement.

The problem is that there is very little other explicit comment on Plummer's teaching from his students. However, the Quaker physician Dr John Fothergill published a short account of the life of his friend (and fellow student) Dr Alexander Russell, which contains a description of Plummer's teaching:

> Plummer is no more; he knew chemistry well; laborious, attentive, and exact, had not a native diffidence veiled his talents as a praelector, he would have been among the foremost in his pupils' esteem. Such was the gentleness of his nature, such his universal knowledge, that in any disputed point in science, the great Maclaurin always appealed to him as to a living library; and yet so great was his modesty, that he spoke to young audiences upon a point he was perfectly master of, not without hesitation.[4]

Even allowing for the notoriously generous and kindly Fothergill's natural wish to not speak ill of the dead, we can glean from this many positive things. Oliver Goldsmith, who attended some medical classes in Edinburgh, wrote that 'Plumer [sic] Professor of Chymistry understands his busines [sic] well but delivers himself so ill that he is but little regarded.'[5]

There is some general agreement which emerges: Plummer apparently was a competent chemist, but a poor lecturer, who was not held particularly high in his students' esteem. Neither Fothergill nor Goldsmith complain about the content of Plummer's course; their grumbles were more about the delivery. This demonstrates a characteristic attitude amongst students of the time; comments by students on their lectures and lecturers suggest that they most admired those professors with impressive and confident delivery, displaying their learning easily, without reading from notes or textbooks. It appears that by this time the lecturer's role had changed from the traditional task of dictating from authoritative texts, perhaps with a little commentary thrown in.[6] Perhaps more importantly, students' expectations had also changed. Originality was prized over tradition and extemporisation over textual authority. The little that is known about Plummer's lectures suggests that he was perhaps a little closer to the older tradition, drawing his course largely from that of Boerhaave, whose student he had been. An early advertisement for the chemistry course to be offered by Plummer and his colleagues reads: 'On Chymistry, being a complete Course, according to the Method of the celebrated Herman Boerhaave, at Leydon, including all the Chymical Processes in the new Edinburgh Dispensatory.'[7]

Plummer's lectures were advertised as being closely tied in to Boerhaave's established authority. His course was not, it would seem, Plummer's course, but Boerhaave's, transplanted from Leiden to Edinburgh. Bower's 1816 *History of the University of Edinburgh* comments on Plummer's teaching:

He was very assiduous in the prosecution of whatever tended to promote the advancement of physic; and, in imitation of his great master, was a zealous cultivator of experimental chemistry. In his class, he taught the theories which were then generally received; but he spent much more time in exhibiting to the students a variety of useful and amusing processes, which were calculated to be of essential service both to the physician and the surgeon. The science of chemistry was then in its infancy, and possessed but few of the allurements which now accompany the study of it. Dr. Plummer directed the attention of his students to a variety of pharmaceutical preparations that were employed in medicine; pointed out the best methods of obtaining them; explained their chemical properties, and how they ought to be employed in practice. So that a great proportion of his course was employed in teaching Pharmacy.[8]

The sources Bower was using to come to this assessment over 60 years after Plummer's death are not given with references. We know that Plummer did teach a number of later celebrated chemists and natural philosophers including John Roebuck, James Hutton, John Walker and James Keir, and it is likely that Cullen himself attended some of Plummer's chemistry lectures, although this is not certain.[9] It might be inferred that Plummer's teaching was sufficiently interesting to inspire a number of his students to continue their study of the discipline. At the very least, we can say that neither Hutton, Roebuck, Walker nor Keir were actually put off chemistry by Plummer. There seems to be no other description of Plummer's teaching beyond Fothergill and Goldsmith's brief comments. This in itself might be telling. Cullen's students were vociferous in their praise of his teaching, and the near silence on the subject of Plummer's, particularly bearing in mind that he taught at Edinburgh for some three times as long as Cullen, again perhaps implies that his chemistry teaching was, at best, uninspiring.

The lack of contemporary commentary on Plummer's teaching does not entirely explain the low standing of his chemistry teaching amongst modern commentators. Black's opinion has been reinforced not by specific reference to Plummer's pedagogical activities, but rather by the implicit denigration of his endeavours by those who wrote about Cullen, as his successor. For example, Thomas Thomson's 1830 *History of Chemistry* is positively elegiac:

The appearance of Dr. Cullen in the College of Edinburgh constitutes a memorable era in the progress of that celebrated school. Hitherto chemistry being reckoned of little importance, had been attended by very few students. When Cullen began to lecture it became a favourite study, almost all the students flocking to hear him, and the chemical class becoming immediately more numerous than any other in the college, anatomy excepted.[10]

Cullen's students were strident in their admiration of their former teacher, and as many of them went on to teach and publish on chemistry themselves, his reputation spread beyond the walls of Edinburgh and Glasgow universities during his life and posthumously. George Fordyce, one of Cullen's early Edinburgh students who went on to teach chemistry in London for 30 years, told his students in 1786:

> The first Dawn of Science in Chemistry was introduced into it by Dr Cullen & hardly any Improvement has been made in the Science since his time: many facts have been lately found out, & this science continues to be constantly enrich'd with these but the Science of Chemistry as far as it is render'd perfect is entirely owing to him.[11]

Cullen's students had warm feelings towards him, and in their zeal to emphasise his importance to their science, the point is consistently made that before him, darkness lay across the land. Plummer's teaching is thus cast into the obscurity of ignorance, enthusiasm and, worse of all, craft. A note of caution might be sounded. Cullen's status and authority as a chemistry lecturer was reflected onto his students, who claimed a not insubstantial authority of their own by emphasising his novelty and his greatness. On this basis, Cullen's standing was in everyone's interest, except perhaps that of the historical record. There is other evidence that might lead us to be wary of such publicly expressed opinions; letters to and from Cullen, around the time of his appointment to the chair of chemistry at Edinburgh, emphasise what an intensely fraught political affair this was. The city abounded with bruised egos and fractured reputations, and it is abundantly clear that Plummer and Cullen were fairly hostile opponents.

In 1751, Plummer, who had by now been teaching chemistry for over 25 years in Edinburgh, sought to give up teaching, while wishing to retain the chair in chemistry. He endeavoured to pass the teaching duties over to a former student, who unfortunately died before being able to take them on. So Plummer, presumably reluctantly, had to continue teaching. There seems to have been some attempt to lure Cullen away from Glasgow at this point, although not by Plummer, but by Cullen's powerful patrons. Cullen wrote in a letter to Dr William Hunter:

I have some thoughts of acceding to a proposal that was lately made to me for removing to Edinburgh. Dr Plummer, Professor of Chemistry, is a very rich man, has given up practice, and had proposed to give up teaching in favour of Dr. Elliot; but this gentleman died about six weeks ago, and, upon this event, some friends of mine, and along with them some gentlemen concerned in the administration of the Town of Edinburgh, have proposed to use their influence with Dr Plummer to induce him to resign in my favour [. . .] However, Plummer's consent, and some other circumstance are still in doubt, and this, with other reasons, requires the affair to be kept as secret as possible.[12]

It appears from the fact that Cullen remained in Glasgow, and that Plummer did indeed continue teaching at Edinburgh, that Plummer was somewhat selective about precisely whom he wanted to take over his duties. Cullen was apparently not the favoured man. It is tempting to speculate about possible reasons for Plummer's apparent antipathy to Cullen, but we have no evidence on this point. All that can be said with confidence is that Plummer preferred to continue teaching himself rather than allow the role to be passed to Cullen. A further episode in 1753–4, when Cullen submitted a paper to the Edinburgh Philosophical Society ('Some Reflections on the Study of Chemistry, and an Essay towards Ascertaining the Different Species of Salts'),[13] can hardly have improved relations. Cullen made some sharp criticisms of 'the narrow views of the teachers of chemistry' that, in combination with the numerous difficulties that obtained in collecting and presenting to students chemical knowledge that was 'scattered in many different writings and very often amongst other matters of very tedious perusal', made the teaching of chemistry at that time flawed. He went on to offer, as an example of his own planned (and improved) method of teaching, a systematic discussion of the chemical doctrine of salts 'as exact as the present state of chemistry will allow'. The comments on the current state of chemical teaching were perhaps somewhat tactless when viewed from the standpoint of a chemistry lecturer of some 27 years' standing, particularly if Cullen had indeed been taught by Plummer during his time in Edinburgh. In that case the criticism, however generally framed, would probably have been taken personally. A letter from Black to Cullen of January 1754 also suggests that a portion of the paper was understood to contradict some work of Plummer's (which had not been published) on the analysis of pit-coal. Black reports some criticisms of Cullen's style as 'too bold [. . .] too careless and prolix', indicating equivocation designed to dampen Cullen's enthusiasm and enable the paper to be rejected without alluding to the real reasons for the obvious snub.[14] Black's letter has a tone of apologetic embarrassment, which is not to be wondered at, and it is clear that neither side was in any way reconciled to the other.

In 1755, Black reported to Cullen that Plummer, at the age of fifty-seven, had suffered an 'apoplexy, which has terminated in a hemiplegia and total loss of speech'.[15] Plummer obviously could not continue to teach. The medical faculty, taking the usual view that a professorial chair was for life, sought, with the support of Plummer's family, to bring Black in to teach in his stead. However, the Town Council had ultimate juris-diction over the occupant of the chair, and they began to put out feelers to find replacement candidates. Black was the preferred option for the medical faculty ultimately to take over the chair (on the presumption that Plummer would never be able to return to his teaching duties). However, the Town Council overcame the faculty, and Cullen, his powerful patrons pleading his case, won the day. In November 1755, Cullen was appointed joint professor with Plummer by the Town Council.

Cullen's soon-to-be colleagues in the medical faculty, were, as Black reported, far from happy. He wrote:

> I am afraid you will receive but little help or encouragement from the Professors. They all seem to be very much out of humour at the Town Council's having managed this affair with so little ceremony, and as if the College had no sort of concern in the matter. Plummer himself will certainly be highly incensed [. . .][16]

This is the same letter from which the earlier quote disparaging Plummer's teaching is drawn. The obstacles put in Cullen's way by the medical faculty were eventually overcome in March 1756 when both Cullen and Plummer resigned tactically and were re-elected as joint professors of chemistry. The whole furore demonstrates how political a matter the bestowal and transfer of a chair at the University of Edinburgh at this time could be. Although the professors certainly endeavoured to wield as much power as the council, Cullen was ambitious and also able to play the political game extremely successfully. The episode also confirms that Plummer cannot be counted amongst Cullen's admirers.

If Cullen's chemistry teaching is to be compared with Plummer's, two aspects might be considered: what was taught, and how it was taught. There is evidence that Plummer's delivery was not universally admired, although there is no reason to doubt his competence in chemistry itself. His chemical papers delivered to the Philosophical Society were competent Boerhaavian efforts which indicate that he knew and understood his discipline.[17] More detail on how Plummer taught is, however, hard to come by. It does appear that the teaching and demonstration part of the chemistry course all took place on the same site, in the chemical laboratory set up and run by the four founding physicians of the medical faculty. An advertisement of September 1729 gives a little more information, stating:

> These three last parts of medicine [i.e. chemistry, the Practice of Medicine and the Institutes of Medicine] are taught in the Chemical Elaboratory adjoining to the University where all sorts of Chymical Medicines thus publickly prepared are sold to Apothecaries.[18]

This advertisement explicitly refers to the professors' enterprising combination of student teaching with the financially lucrative production and sale of drugs. In 1726, Plummer, Innes, Rutherford and St Clair had entered into a business partnership to prepare and sell medicines from the pharmaceutical laboratory in which they were also teaching students. It has been noted that this business of preparing and selling drugs was extremely profitable to Plummer and his colleagues, and it seems likely that it was this business rather than the fees gained from his chemistry lectures that made Plummer, as Cullen put it, 'a very rich man'.[19] Lecturing duties thus seem to have been combined with these business activities, and this tactic would undoubtedly have lent Plummer's course a pharmaceutical bias. It is interesting to speculate on how the demands of a thriving business might have influenced the course structure and delivery.

There is even less information about the site of Cullen's lectures. Due to the fraught nature of his succession to Plummer's chair, it appears that Plummer's laboratory was not made available to Cullen when he first arrived. There was some talk of Cullen purchasing Plummer's laboratory, but it is unclear whether this actually occurred. An alternative site was available, a laboratory that had been used by Plummer's assistant, James Scott, and it is thought that this is the laboratory that Cullen used.[20] There is also evidence that Cullen extended his teaching for selected students to the less formal surroundings of his own home. A letter from an American student, Samuel Bard, to his father, of 4 February 1764 explains:

> Dr. Cullen has lately entertained me much, by some private lectures he gives to those who attend him for the second year upon what he calls the Chemical Pathology, in which he attempts to prove the presence and necessity of an acid generated in the stomach; and endeavours to account for the assimilation of the aliment, upon more rational principles than the extravagant theories of the Corpuscularians and older chemists. What I chiefly admire is the manner of them; we are convened at his own house, once or twice a week, where, after lecturing for one hour, we spend another in an easy conversation upon the subject of the last evening lecture, and every one is encouraged to make his remarks or objections with the greatest freedom.[21]

Bard was apparently keen to give a picture to his family at home in New York of his life in Edinburgh, and an earlier letter is similarly informative on Cullen's more everyday chemistry lectures:

> [Cullen] is a very good speaker, and very eminent in his profession; lectures in English, in a clear, nervous style, and with a natural strong tone of voice. He has a new way of examining his pupils in his lecture room; and as I was recommended to his notice, he did me the honour, this winter, to commence with me; from which I would as lieve have been excused, for I was not a little confused to be thus questioned before above a hundred students, who all had their eyes fixed upon me, to hear my answers [...].[22]

Cullen, it would seem, took a novel approach to his pedagogical practice. Rather than simply 'lecturing' his students, he tested their understanding by questioning them before their fellows, making the pedagogical process less of a one-way street. Both his new way of 'examining' students in the formal lectures and the additional, extra-curricular pedagogical sessions at his home, where new ideas and theories were discussed and debated in an informal atmosphere, encouraged students to feel part of the process of knowledge generation rather than passive recipients of information.

For more information on how a science like chemistry, with both practical and theoretical knowledge to be disseminated, was taught, student lecture notes can be consulted. Archives all over the world, but particularly in Scotland, contain sets of notes from the chemistry and medical lectures of both Cullen and Black.[23] These notes can be of various kinds. Few surviving sets are those written by students while in attendance at the lecture. Some were written out neatly afterwards, but many seem to be written by professional scribes. Silas Neville records late nights spent 'extending his notes taken at the chemical and anatomical lectures'.[24] Some students appear to have gathered after lectures to compare notes and thereby produce a perfect copy of what had been said. One set of notes brought back to Philadelphia from Edinburgh by Thomas Parke is annotated as the production of 'Thomas Parke and Co'.[25] Students also copied notes they had borrowed from their fellows. Students may have had sight of sets of notes even before attending the university; Benjamin Rush wrote to John Morgan: 'I am now more fully convinced than ever how much sleep you must have sacrificed in transcribing those volumes of learning you carried with you to America.'[26]

Rush found that some students had learned shorthand in advance of their attendance so as to enable the easy recording of every word spoken by their professors.[27] Once the notes were perfected, they were copied; by the middle of the eighteenth century there was a thriving industry in the production of chemical and medical lecture notes, which enabled even those who could not attend the university to pick up some gems of knowledge from the famous professors. John Elliot, an apothecary living in London in the 1780s, recorded his pleasure at having been able to study a set of Black's lecture notes belonging to a friend, although he complained that 'he only wished they had been more perfect'[28].

This frenzied activity amongst students to ensure that their lectures were preserved for the future was prompted only in part by a desire to have clear notes to refer to for their own benefit. As we have seen from Rush's letter, students returned home to their own communities and clear sets of notes enabled them to pass on the knowledge gained to young friends and relations. Other students, no doubt, saw an opportunity for financial gain. This was perhaps prompted by the success (for someone, at least) of an unauthorised textbook based on Boerhaave's lectures. It was this event that provoked Boerhaave, in righteous fury, to put together his own chemical textbook – as he said, to correct the errors that were attributed to him in the unauthorised version. Similarly, when a set of student notes taken from Cullen's lectures on materia medica was published without his authority or approval, the imperfections of the initial publication prompted Cullen to compile his own book, rewritten and enlarged. And a similar thing was to happen to Black, who was, it would seem, reluctant to compile his own textbook; but even when the anonymous *Enquiry into the Effects of Heat and Mixture* was published in 1770, this did not persuade Black to issue his own authorised version.[29]

Lecture notes, though useful as the best record we have of what was taught to students in the eighteenth century, should be assessed with care, as in the process of taking them down, rewriting them and copying them, errors and misunderstandings could creep in. In addition, of course, the inattentive student could miss essential points or, if lazy or prone to illness, whole lectures (although the practice of 'group editing' mentioned above was obviously a practical way to solve this problem). A twentieth-century study of student note-taking found that the percentage of 'information units' students recorded in their notes varied across different sections of the lecture between 33% and 5%, and averaged 21% overall.[30] So modern students only recorded one in five of the information units presented. Clearly we cannot assume that, had the same study taken place 200 years earlier with quite different teaching, lecturing and learning practices in play, the results would have been the same. But we should perhaps take this as a cautionary tale when considering what historians can learn from lecture notes; they are likely to be mere shadows of the lectures themselves.

Having said all that, of course, they remain the best source we have for what was actually said, done, witnessed and learned in the lecture hall. And the simple fact that for Cullen there are so many sets still extant, in addition to the fact that Glasgow University Library also holds some of his own notes, means that some account of what and how he taught can be put together. However, the same cannot be said for Plummer. There are only three sets of notes that are believed to be notes from Plummer's lectures. The first is a four-volume set which describes 213 pharmaceutical preparations in Latin, followed by two lectures in English on essential oils and fermentation.[31] The second set is in English, undated, and consists of 16 small pages headed 'The Order of the Processes in Chemistry', describing 49 processes in an increasingly crabbed and cramped hand.[32] It is perhaps relevant to note here that the unauthorised edition of Boerhaave's chemistry lectures was published in Latin before Plummer began teaching, and in English translation shortly afterwards, and that the authorised version appeared in both Latin and English only a few years later. The processes follow the order as laid out in Volume 2 of the authorised version, the volume concerned with the Practice of Chemistry. Hence the first process encountered is a 'water, exhaled in the form of vapour from a recent Plant e.g. Rosemary, by a heat equal to that of the sun in summer'. This is as simple a process as could be imagined, requiring only water and a very gentle heat. Boerhaave's first process was the same; also on rosemary. The correspondences are so close that it might be thought that the notes were simply a student's abridgement of Boerhaave – were it not for the fact that on page 11 the student departs from the description of the processes, just after having described process number 23, the distillation of an essential oil, to say, 'Here follow the distillation of a great many oyls in as many processes which Dr. Plummer spoke of in general so that that which is next in his order is the 32^{d} of Boerhaaves'. From the limited information we have, then, it would appear that Plummer simply marched his students through Boerhaave's course, missing out a few processes which appeared unnecessarily repetitive.

These two sets of lecture notes, and probably Bower's comments, have given rise to what has become a common theme in discussions of Plummer's chemistry teaching; that he taught the minimum of theory and primarily concentrated on pharmaceutical processes and preparations. Today, Plummer's teaching is disparaged on these very grounds. However, there is no good reason to expect more: Plummer was one of the four founding professors of the Edinburgh Medical School. George Drummond, the Provost of Edinburgh, perhaps explained the situation most clearly in a letter to William Cullen of 1756. He wrote:

> I am quite sensible that there is a wide difference between the situation the Professors of Medicine were in at the time they were established in the University and yours. At that time, we were only making a trial, and were somewhat uncertain about its success;

and yet they thought themselves favoured by the Town-Council in giving them a preference in our choice to another set who petitioned us to appoint them [...] Every body is now sensible that the medical college is an immense benefit to the community, and must continue to be so as long as the Town-Council do justice to the Town, by filling up vacancies when they happen with professors of established reputation. Such is the present case. You don't court us to come to you, we court you to come to us.[33]

As Drummond said, Cullen's position in 1756 was quite different from that of Plummer 30 years earlier. Plummer was effectively on trial, along with his colleagues. They were setting up a new venture, a medical school, and as such the chemistry that they were supposed to be teaching was specifically intended to be used by physicians to enable them to prescribe and possibly on occasion prepare medicines. Thus, Plummer taught medical chemistry, although his personal interests were wider. To disparage Plummer's chemical teaching purely on the grounds that he 'only' taught pharmaceutical chemistry is to take an ahistorical view; to charge Plummer with failing to be Cullen, in a sense, as if he missed an obvious opportunity to extend his science beyond the bounds of pharmacy – failing, in effect, where Cullen later succeeded.

The lecture notes we have do, nevertheless, support the view that Plummer's chemistry lectures were predominantly devoted to pharmaceutical preparations and processes. Such theory as these lectures contain is of the most general kind and is included more by implication than as a result of any specific discussion of chemical theory. These are the only two sets of lecture notes that are currently available. There is, however, somewhere, a third set, which, if it could be located, might give us much more information about Plummer's chemical pedagogy. Up to the early 1980s, this set of lecture notes was held at the Wellcome Library in London, as MS 3923. However, as the Wellcome catalogue now notes, since 1983 this set of notes has been missing. It is known from the Wellcome catalogue entry that although it was only a single volume, it was a fairly substantial set of notes at some 356 pages; and further, it is known that Plummer's name is spelt as 'Plumber' on the binding. There is reason to believe that the contents of these lectures would, if found, give us a great deal more insight into Plummer's pedagogy. The 1969 doctoral thesis of John Crellin, a postgraduate at the University of London, includes some references to this lost set of Plummer lecture notes. Crellin appears to have been one of the few scholars who spent any substantial time examining the manuscript, and who included his thoughts in his thesis. He firmly dates the manuscript as 1746, indicating that it was dated as such. However, only a few comments are offered on the contents, which have perhaps gained greater significance in the years since he originally made them, in the light of the loss of the original manuscript. Crellin's comments on Plummer's lectures is the best evidence we now have for the content of these lectures, and this in itself may prompt some reflection on the nature of the constraints placed on our use of such sources and how legitimate inferences can be drawn from them.

Crellin asserts that Plummer was conscientious and took some care to revise his course in the light of new information, based on the fact that the lectures included an account of George Martine's work on thermometers, which had been published in 1740, only six years earlier. He claims that Plummer was keen to stress the general importance of the discipline, and quoted Thomas Sprat in saying that the arts will only arrive at 'perfection when ye inventors [. . .] have mechanics hands or ye mechanics philosophers heads'. Crellin also indicates that Plummer took some care in formulating rules to plan the experimental part of his course, offering the following quotation, which, as it is all the remaining text of the lost manuscript, is given in full:

> I shall lay down 3 rules wch if duly observed will produce a good system of chymical experiments. The first rule is yt each kingdom of bodies should be considered distinctly and treated separately. There are some particular properties wch agree to all ye bodies of one class & have elements diff^r to those of diff^r classes. As ye subjects of one kingdom have some properties belonging to all, all ye lower class have their particular properties. The 2nd rule is yt we begin our experiments on ye general class wch can be sufficiently known without other bodies. It is indifferent with which we begin if we observe ye first rules. The neglect of this 2nd rule renders chymistry very foolish. Lemery is very defective in this order; he thought yt minerals are more simple yn plants or animals; but it is plain yt he can make no experiment without employing other bodies whose history is unknown to a beginner.[34]

Plummer's claim to have ordered his processes carefully seems slightly disingenuous in the light of the Wellcome manuscript, where he followed Boerhaave's order so closely. Indeed, at least some of his 'rule' is a rephrasing of Boerhaave's statement in the preface to the second volume of his *Elements of Chemistry* that 'I have always shewed whatever ought to proceed, before I proceed to what ought to follow.'[35] This seems to confirm the notion that Plummer was simply echoing Boerhaave. However, Crellin argues that, contrary to what has been seen elsewhere, Plummer did consider and critically assess chemical theory in his lectures. The quotation he gives is indeed critical of Lemery's work, although it would seem to be criticism of the ordering of his course, rather than of any chemical theory as such. However, without taking Crellin's views into account, on the evidence of this brief quote out of the total 356 pages of the missing manuscript, it can only be said that even in his theoretical justification of the ordering of his course, he was not notably departing from Boerhaave's principles, and there is still no evidence that he taught anything outside the boundaries of pharmacy.

Besides the shorter set of notes mentioned above, Glasgow University Library also holds a companion manuscript of three pages of the same size, in the same hand – and from the records, it would appear to be the work of the same student – titled 'Chemical Experiments I tryed'.[36] This appears to record the attempts of the student to follow Plummer's instructions by attempting some of the processes described in the notes. He carefully describes his attempts (although some time after the fact – see below) and their results and indicates which numbered process each one is. These processes match the ones in the lecture notes, although unfortunately it seems that his results did not accord particularly well with what was expected. He notes of one process, the simplest of all, that 'it proceeded vastly slowly' so he turned up the fire a little; the resulting substance, he states disappointedly, '5 days after, begins to turn a little thickish tho it has been nicely stopt'. Conversely, another process designed to produce a 'gelly', turned out 'not at all thick but a great deal browner'. It seems likely that much of his trouble stemmed from the difficulty of attaining the appropriate level of heat, and from a certain vagueness about timing. He seems to have got bored quickly and on occasion left his process to continue overnight, and in other cases, terminated it early. The last process he records was the burning of the plant ashes, 'yet they were not white but a light brown'; when they were dissolved in water, boiled and evaporated he got 'near half a drachm of salt [...] it was brown, oily, pungent & acrid & soon moistened when exposed to the air'. It is clear that this was not the hoped-for result, and we might wonder what, if anything, Plummer offered his students that was not available in a copy of Boerhaave's course. From this student's notes, and his account of his experimental efforts, it does not appear that there was very much. The tacit knowledge required to complete the processes successfully does not seem to have been communicated effectively, and the student appears to have been left somewhat at a loss, perhaps compounded by a lack of theoretical grounding.

This manuscript implies that the unknown student had found somewhere to attempt his own chemical experiments, but it is far from clear where and what facilities were available to Plummer's students for them to develop their own practical skills. It is also unclear from Plummer's lectures whether the various processes he ran through were demonstrated during the lectures. It seems unlikely, given that many of the processes required heating for hours, if not days. Bower claims that Plummer exhibited 'useful and amusing processes' (see above), but sadly we have no information about these.[37] Most students lived in rented rooms in the town, and although it is possible that some enterprising students might have made use of portable furnaces to make certain trials at home, we have no evidence confirming this. The fact that Plummer's laboratory was used for the commercial production of drugs as well as lectures might suggest that there was only limited time and space available for the use of students. At Apothecaries' Hall in London, a 1753 proposal to inaugurate lectures to take place in the laboratory was rejected as it was feared that such activities might inhibit the laboratory's drug production.[38] A commercially successful laboratory would have been a busy and possibly chaotic place, and how much opportunity there was for students to get hands-on experience of the practice of chemistry is unclear. It is possible that selected students provided a cheap, or even free, source of labour for the drug production enterprise; although, as the unknown student's notes of his trials indicates, quality control would have been an ongoing problem.

From the first days of his lectures at Glasgow, Cullen regarded practical experience as an essential part of the student learning process. He famously chided his early students for their apparent reluctance to make use of the laboratory that was placed at their disposal.[39] In Edinburgh, although there is some doubt about where Cullen taught, the facilities certainly included a laboratory, and it is inconceivable that he would not have made arrangements for students to use this, with the advantage that they would not have had to compete with commerce for their access.[40]

Where Plummer apparently perceived chemistry as primarily concerned with pharmacy, Cullen took a broader view. Some notes that are believed to be Cullen's own from the earliest years of his chemistry lectures at Glasgow comment:

> Chemistry is still a new study here & I find that very absurd false narrow & confused notions prevail with regard to it [. . .] those Tradesmen who in great Cities take to themselves the name of Chemists are apt to think that the sole business of the art is the preparation of Medicines & indeed it hath been too much confined to this purpose which is a very small branch of it [. . .][41]

The criticism of the current state of chemical pedagogy was not confined to the paper sent to the Philosophical Society of Edinburgh, but was, it appears, also voiced in his lectures:

> it was particularly to be expected, that, since it has been taught in universities, the difficulties in this study should have been in some measure removed, that the art should have been put into some form, and a system of it attempted – the scattered facts collected and arranged in a proper order. But this has not yet been done; chemistry has been taught but upon a very narrow plan. The teachers of it have still confined themselves to the purposes of pharmacy and medicine, and that comprehends a small branch of chemistry; and even that, by being a single branch, could not by itself be tolerably explained.[42]

Cullen approached the first years of his chemistry teaching determined to develop a course that satisfied his own high expectations, and offered what he considered to be a proper grounding in what he termed 'philosophical chemistry'. His course addressed the problems he perceived, offering a system that would allow the aforementioned 'scattered facts' to be arranged in a proper order and offer a level of theory to underpin that order and explain the processes and practices of the discipline.

Over the years he taught at Glasgow, Cullen developed his course, and by the time he reached Edinburgh, its form was established and the barbed comments about his predecessors appear to have been largely dropped. By this time, his status was assured, and the mood was conciliatory rather than disparaging. His practice, followed in most of the surviving sets of lecture notes, was to begin with a history of chemistry, then cover what he termed the objects of chemistry – those bodies that chemists manipulated to produce new substances or bodies with particular qualities. These bodies he classified into salts, inflammables, metallic, earthy, watery and, in later years, aerial bodies, on the basis of a group of simply evaluated qualities such as flammability, solubility, taste, form, etc. Included in this section was a discussion of elements, principles and compounds, about which he took a pragmatic view:

> We therefore apply the term Element to signify any very simple sort of Matter which enters into the Composition of several other Bodies; that is Chemical Principles [. . .] we use the terms Mixt and Compound indiscriminately to signify a Body combined of any matters possessing different qualities.[43]

He continued with a description of the operations of chemistry. It is here that the most fundamental aspect of his course was expounded. All chemical change, according to Cullen, was accomplished by the combination or separation of particular bodies to produce new ones with different properties. This, the chemist accomplished by the judicious use either of fire or of the selective natural inclination in all bodies to combine, termed by Cullen 'elective attraction'.

All operations in chemistry could thus be ordered according to whether they produced combination or separation and whether they made use of fire or elective attraction – or both – to accomplish the desired result. Thus, as Cullen said, 'all the several operations of chemistry may be referred to solution, fusion or exhalation. But that the manner be triple, the power of operating is but double, viz. attraction and the action of fire.'[44] Fusion and exhalation were actions of fire, used to separate substances due to their different volatilities at different temperatures, while solution allowed the elective attractions between different bodies to act.

Elective attractions or affinities, and the ability to manipulate these natural attributes of all substances, were at the heart of Cullen's chemical system. Cullen's affinity theory offered a demarcation principle to distinguish chemistry and its methods from the mechanical philosophy. Presented simply, but with the potential for fiendish complexity, his affinity theory would allow his students to at least make a reasonable conjecture as to how they might use known bodies to produce new, as yet unknown ones with desirable properties. He developed innovative diagrams that spread across the chemical world, depicting the play of competing affinities within a chemical mixture.[45] Affinity would enable a complex process to be explained and understood, and allow the student who could master his system to predict the results of unknown combinations. Cullen's lectures were thus the scene of knowledge generation as well as dissemination; in utilising affinity theory as the organising principle behind his course, he assigned it an explanatory and predictive role that it had not yet held. His students would carry this forward from his lectures into the wider world. In this case, pedagogical innovation would change the discipline beyond the university walls.

The second half of his course consisted of what he called a 'history of bodies'; that is, an account of the various substances encountered in chemistry, ordered, as he said, 'after the manner of the Botanists into Classes, Orders, Genera, Species, Varieties'.[46] Although Cullen did recommend some books to his students, his course, unlike Plummer's, did not follow any one authoritative text. Indeed, he took care to prepare his own syllabus and have it printed for his students' benefit at both Glasgow and Edinburgh. One student, Charles Blagden, began his notes of Cullen's 1766 lectures with regular references to see particular pages of a syllabus extending to at least 32 pages. It is possibly significant that, while students were assiduous in preserving their lecture notes, in the vast majority of cases the accompanying syllabi seem to have been lost. It might be expected that more of these documents would have survived than the manuscript lecture notes, as most were printed and sold to potential students; but other than a single copy of a very early printed syllabus amongst the Cullen papers at Glasgow University, none seems to have done.[47] This might, of course, be more in consequence of the relative importance assigned to manuscripts in comparison with printed ephemera in the minds of archivists and collectors as opposed to the intentions of the original students. Similarly, affinity tables were also printed and made available to students, presumably at a small fee – one of these does, however, survive in the National Library of Wales.[48]

Cullen's lectures were not simply auditory experiences, but were enhanced with visual and other sensory refinements. We know from a number of student lecture notes that he included a variety of simple demonstrations in his lectures – these seem to have been predominantly solution and precipitation chemistry, and were necessarily fairly swift, with the chemical change being clearly sensible through colour changes, appearance and disappearance of bodies, fumes, smells, etc. The production of fumes when vitriolic acid was poured on to nitre was one simple demonstration described in a late set of lectures. The action of heat and fire, however, was not neglected. One set of lectures states 'a melting or wind furnace was now shown in action'.[49] We also know that Cullen pinned up tables for his students to copy into their notebooks. A table showing the characteristic qualities of acid, alkali and neutral salts was shown, as was an affinity table. The content of these tables was justified by carefully chosen demonstrations that showed in brief how they were produced by generalising from series of individual experiments.

In conclusion, any discussion of Plummer's chemistry lectures seems to involve the constant reiteration of the phrase 'there is no evidence', which is frustrating, but hardly an uncommon occurrence for historians. It is when this paucity of information is contrasted with the relative abundance of evidence about Cullen's teaching that it becomes suggestive. It might be speculated that the very fact that there is so little is itself of some significance. If Plummer's students had found his lectures as valuable as Cullen's did his, might they not unreasonably be expected to have left more information about them? But it could also be argued, with some grounds, that note-taking in lectures was a growing trend throughout the eighteenth century, and that students in general simply took fewer notes prior to 1750. It is certainly true that the majority of sets of notes of Cullen's lectures still extant date from his Edinburgh days, and that very few sets from Glasgow exist. It is possible that the lack of notes of Plummer's lectures is due to the fact that the practice was not so widespread as it became after his death. On this basis, his reputation might be tarnished unwarrantedly, as a secondary result of the way in which lecturing and learning practices developed. Many well-known lecturers in chemistry who were also lecturing, for the most part privately, are also not represented in the archives. We have no notes of William Lewis's lectures, or of Peter Shaw's, for example; Plummer would appear to be in excellent company. On the other hand, there are lecture notes in collections worldwide that date from the very period Plummer was teaching; a quick search will locate manuscript notes of Boerhaave's lectures, along with those of Alexander Monro *primus*, Charles Alston, and a host of others from the first half of the eighteenth century and plenty from before. Some students, it seems, were certainly taking notes in their lectures at this time, so while it seems likely that there was a growing trend in the years after Cullen took over Plummer's chair at Edinburgh for students to take, copy and keep their notes, this will not serve to explain the situation entirely.

Another possibility, stemming from the fact that the few sources we have suggest that Plummer essentially taught the second volume of Boerhaave's *Elements of Chemistry*, is that Plummer followed Boerhaave's course so closely that his students for the most part would have only required a copy of Boerhaave's work, when such was available. Is it reasonable, too, to claim on the basis of the little information we have that Plummer didn't teach theory? The fact that he taught Volume Two of Boerhaave's course means it is entirely feasible that he also covered, although perhaps not in any great depth, the contents of Volume One, dealing with the theory of chemistry. Speculating wildly, it could be argued that the reason there are no lecture notes of the first volume is because most students merely annotated their copies of Boerhaave. This is of course hypothesising far beyond the bounds of the meagre evidence, but I would suggest that Plummer's pedagogy was by no means as worthless as Black claimed, although it was probably never particularly original. It is tempting, when there is very little evidence, to grant a great deal more credence to the little that remains than is historically defensible. It is entirely justifiable, when using the scanty sources available, to make claims about what Plummer did teach; but using them to make claims about what he failed to teach (and how he failed to teach it) is difficult to defend.

It can be argued with perhaps a little more conviction that the reason so many sets of notes remain for Cullen's lectures is a consequence of Cullen's determination to offer an original course, without much reference to textual authority. This was prompted in part by his desire to bring his discipline out of the shadow of the medical school and emphasise its utility throughout society. In this regard, and in the way he structured and organised his course content, in how he sought to settle the doctrine of affinity at the centre of his discipline, developing useful tools to enable his students to think like chemists rather than just prepare medicines, Cullen's chemical pedagogy was indeed novel. The way he taught was certainly original; allowing the student voices to be heard in his lectures, as well as offering extra-curricular events and courses to encourage students to discuss and debate new ideas and even to form their own speculative theories. Cullen's students were taught to think critically about their discipline, even elevated to the status of assistants in the knowledge-generation enterprise that his teaching became. He valued his students' opinions and thoughts, and gained their respect and admiration in return. His innovative teaching became famous across Europe and America, and as a result, students from both continents flocked to attend his courses. The number of lecture notes surviving is also related, of course, to the numbers of students who took Cullen's course each year, and as Bard's letter to his father testifies, this was large. Plummer's courses, however, are unlikely to have been ill-attended, at least in the latter years of his career. Drummond's letter to Cullen of 1756 implies that the medical school had been a success, measured, most likely, in student numbers, and the figures we have do testify to this. In the five years from 1746–51 (when Plummer was teaching), 18 students graduated MD out of a total of 364 matriculating. While only a small proportion graduated, the number matriculating was not inconsiderable. A little later, from 1766–71 (when Cullen and Black had taken over), a total of 655 students matriculated (a notable increase on the earlier years), but only 17 graduated MD.[50] While the rise in students matriculating was notable, it was not such as to account for the disparity in itself.

However, even with all the evidence there is to draw on, there are still a number of questions that remain unanswered about Cullen's pedagogy. Bard's letters, particularly when considered in conjunction with the evidence of the lecture notes, raise some intriguing questions. What kind and arrangement of space allowed Cullen to lecture to 'above a hundred students' effectively, so that they could all hear him speak, witness demonstrations and appreciate the sometimes subtle results, see the action and effects of furnaces, take copious notes and copy the tables he 'pinned up'? If the complications of taking any kind of written notes at the time are taken into account – requiring paper, ink, inkwells, quill pens (which could be alarmingly temperamental), penknives for sharpening said pens, sand or pounce for blotting, etc. – it becomes clear that we know much less about the practical, physical side of the teaching and learning process in the eighteen century than about the content of chemistry courses. There is certainly more information about *what* was taught than *how* it was taught – or, what is perhaps of more interest, how it was learned. The history of student learning practices is a largely unexplored field at present. It is to be hoped that in the future, despite the difficulties outlined above in mining the limited array of sources available, historians will rise to the challenge, and in time will be able to reveal the practical secrets of the lecture hall.

Notes and References

1 Letter, Joseph Black to William Cullen, 22 November 1755, cited in Anderson, R. G.W. and J. Jones (eds), *The Correspondence of Joseph Black* (Farnham: Ashgate, 2012), vol. 1, p. 130.

2 Thomson, J., *An account of the life, lectures and writings of William Cullen* (Edinburgh: Blackwood, 1832), p. 92.

3 Anderson and Jones, *Correspondence of Joseph Black*, vol. 1, p. 125, note 3.

4 Fothergill, J., *An essay on the character of the late Alexander Russel . . . Read before the Society of Physicians, the 2d of October, 1769* (London, 1770), p. 5.

5 Oliver Goldsmith to the Rev. Thomas Contarine, 1753, in Goldsmith, O., *The Collected Letters of Oliver Goldsmith* (Cambridge: Cambridge University Press, 1969), p. 6.

6 Friesen, N., 'The lecture as a Transmedial Pedagogical Form: A Historical Analysis', *Educational Researcher* 40 (2011), pp. 95–102.

7 *Caledonian Mercury*, October 1725, quoted in Doyle, W.P., *Andrew Plummer M.D. (1697–1756)* (Edinburgh: Scotland's Cultural Heritage, 1982).

8 Bower, A., *History the University of Edinburgh*, vol. 2 (Edinburgh: Oliphant, Waugh and Innes, 1817), pp. 215–16.

9 See Doyle, *Andrew Plummer M.D.*, ref. 7. Cullen was certainly in Edinburgh during the period that Plummer was lecturing in chemistry, and it seems inconceivable that he would not have attended at least some of Plummer's lectures, given his clear interest in the subject. W.P.D. Wightman ('William Cullen and the Teaching of Chemistry', *Annals of Science* 11 (1955), pp. 154–165) claims that he did, but A.L. Donovan (*Philosophical Chemistry in the Scottish Enlightenment* (Edinburgh: Edinburgh University Press, 1975)) says that there is no positive evidence to confirm this.

10 Thomson, *An account of the life*, p. 307.

11 Fordyce, G., *Lectures on Chemistry* (1786), Royal College of Physicians, London MS 146–148, Lecture 1st.

12 Thomson, *An account of the life*, p. 82.

13 This paper was never published, but has been reconstructed from surviving manuscript fragments and can be found in Dobbin, L., 'A Cullen Chemical Manuscript of 1753', *Annals of Science* 1 (1936), pp. 138–156.

14 Letter, Joseph Black to William Cullen, Edinburgh, n.d. January 1754, in Anderson and Jones, *Correspondence of Joseph Black*, vol. 1, p. 115.

15 Letter, Joseph Black to William Cullen, Edinburgh, n.d. (1755), in Anderson and Jones, *Correspondence of Joseph Black*, vol. 1, p. 125.

16 Letter, Joseph Black to William Cullen, Edinburgh, 22 November 1755, in Anderson and Jones, *Correspondence of Joseph Black*, vol. 1, p. 130.

17 John Power, in this volume, is particularly enlightening on the influence of Boerhaave on his students, amongst whom Plummer is included.

18 *Caledonian Mercury*, 29 September 1729, p. 4.

19 See Anderson, R.G.W., *The Playfair Collection and the Teaching of Chemistry at the University of Edinburgh, 1713–1858* (Edinburgh: Royal Scottish Museum, 1978), p. 10.

20 Ibid.

21 McVickar, J., *A Domestick Narrative of the Life of Samuel Bard, M.D., L.L.D., Late President of the College of Physicians and Surgeons in the University of the State of New York* (New York: Columbia College, 1822), p. 55.

22 Letter, Samuel Bard to his parents, 5 December 1762, quoted in McVickar, *A Domestic Narrative*, p. 35.

23 Edinburgh and Glasgow University Libraries hold examples of lecture notes from Cullen's and Black's lectures, as do the Royal College of Physicians in Edinburgh, the Wellcome Trust Library in London, Aberdeen University Library, the library of University College London and various libraries in America. Wightman, 'William Cullen' and Wightman, W.P.D., 'William Cullen and the Teaching of Chemistry – II', *Annals of Science* 12 (1956), pp. 192–205 discuss the Cullen papers at Glasgow in depth. See also Cole, William A., 'Manuscripts of Joseph Black's Lectures on Chemistry' in A.D.C. Simpson, 'Joseph Black 1728–1799: A Commemorative Symposium' (Edinburgh: Royal Scottish Museum, 1982), pp. 53–69.

24 Entry for 12 November 1771, Cozens-Hardy, B. (ed.), *Diary of Sylas Neville 1767–1788* (London and New York: Oxford University Press, 1950), p. 140.

25 Bell, Whitfield J., Jr, 'Thomas Parke's Student Life in England and Scotland 1771–1773', *The Pennsylvania Magazine of History and Biography* 7 (1951), pp. 237–59 (on p. 250).

26 Quoted in Freeman Hawke, D., *Benjamin Rush: Revolutionary Gadfly* (Indianapolis, IN: Bobbs-Merrill, 1971), p. 47.

27 Indeed, there is a set of lecture notes held by the Wellcome Library, London, MS 5281, apparently taken from the chemistry lectures of George Fordyce, one of Cullen's students, which is predominantly (aside from headings and index) in shorthand.

28 Elliot, J., *Philosophical Observations on the Senses of Vision and Hearing; to which are added, a Treatise on Harmonic Sounds, and an Essay on Combustion and Animal Heat. By J. Elliott, Apothecary* (London: J. Murray, 1780), pp. 122–3.

29 Anon, *An Enquiry into the General Effects of Heat; with Observations on*

the *Theories of Mixture. In Two Parts: Illustrated with a Variety of Experiments, Tending to Explain and Deduce from Principles, Some of the Most Common Appearances in Nature. With an Appendix on the Form and Use of the Principal Vessels Containing the Subjects on which the Effects of Heat and Mixture are to be Produced* (London: J. Nourse, 1770).

30 Hartley, J. and A. Cameron, *Some Observations on the Efficiency of Lecturing, Educational Review* 20 (1967), pp. 30–7.

31 Plummer, A., *Opera Chemica: A. Plummer,* Royal College of Physicians, Edinburgh, MS M8.

32 Anon., 'The Order of the Processes in Chemistry' (undated, perhaps 1736), Glasgow University Library, MS H487.

33 Letter, George Drummond to William Cullen, 3 February 1756, quoted in Thomson, *An account of the life*, pp. 94–6.

34 Wellcome MS 3923 (previously MS 119), quoted in Crellin, J.K. *The Development of Chemistry in Britain through Medicine and Pharmacy, 1700–1850*, University of London PhD thesis (1969), pp. 200–1.

35 Boerhaave, H., 'Preface', in *Elements of chemistry: being the annual lectures of Herman Boerhaave, M.D. formerly professor of chemistry and botany, and at present, professor of physick in the University of Leyden*, trans. Timothy Dallowe, vol. 2 (London: J. and J. Pemberton, J. Clarke, A. Millar and J. Gray, 1735).

36 Anon., 'Chemical Experiments I Tryed', Glasgow University Library, MS H486 (1736).

37 Bower, A., *History of the University of Edinburgh*, vol. 2 (Edinburgh: Oliphant, Waugh and Innes, 1817), pp. 215–16.

38 Simmons, A., 'Stills, Status, Stocks and Science: The Laboratories at Apothecaries' Hall in the Nineteenth Century', *Ambix* 61 (2014), pp. 141–161 (on p. 149, fn. 43).

39 Glasgow University Library MS, quoted in Wightman, 'William Cullen – II', p. 196.

40 See Anderson, *Playfair Collection*, pp. 12–14.

41 Cullen, W., *General Idea of Chemistry*, Glasgow University Library, Cullen Papers, MS Cullen 1059.

42 Thomson, *An account of the life*, p. 40.

43 Blagden, C., *Lectures on Chemistry given by William Cullen* (1766), Wellcome Library, London, MS Cullen Papers, MS1922, lecture 18.

44 Ibid., lecture 24.

45 Crosland, M.P., 'The use of diagrams as chemical 'equations' in the lecture notes of William Cullen and Joseph Black', *Annals of Science* 15 (1959), pp. 75–90.

46 Blagden, C., *Lectures on Chemistry*, lecture 10.

47 Cullen, W., *The plan of a course of chemical lectures and experiments to be given in the College in Glasgow during the session MDCCXLVIII* (1748), Glasgow University Library, Cullen Papers, MS Cullen 1069.

48 Cullen, W., *Dr Cullen's Table of Elective Attractions*, National Library of Wales, MS 2568E.

49 Blagden, C., *Lectures on Chemistry*, lecture 43.

50 Swinbank, P., 'Experimental Science in the University of Glasgow at the Time of Joseph Black', in Simpson, *Joseph Black: A Commemorative Symposium*, pp. 26–7.

FIVE

'The Most Perfect Liberty': Professors and Students in the Age of the Chemical Revolution

JOHN R.R. CHRISTIE

Joseph Black was Scotland's pre-eminent chemist of the eighteenth century. A student of William Cullen's at Glasgow, his reputation was founded upon his early-career researches into the chemistry of mild and caustic alkalis and the related function of 'fixed air' (carbon dioxide), and into specific and latent heats. After teaching chemistry for a decade at the University of Glasgow, he took up the Edinburgh professorship of chemistry in the medical school, succeeding William Cullen in 1766, and there he taught an annual course of chemistry for the next three decades. Black's fundamental chemical research is well known and well studied, and he may also be acknowledged as an influential teacher of chemistry, with pupils such as his successor at Edinburgh, Thomas Charles Hope; Benjamin Rush, America's first professor of chemistry; Thomas Beddoes, founder of the Bristol Pneumatic Institute (after teaching at Oxford); and James Smithson, whose substantial legacy went to the founding of the Smithsonian Institution.

Before moving into the substance of this essay, some account of Black's chemical teaching, drawing on contemporary descriptions and reminiscence, may function as useful background to what follows. Not all are exclusively laudatory: 'good language and very instructive, but I blush for his delivery', wrote one American student in the 1770s.[1] Sylas Neville noted, 'Dr. Black is without an equal in chemistry,' but remarked two years later, 'I cannot help thinking that abridging his [Black's] proper course after beginning another at an early hour to a set of lawyers &c has not the best appearance.'[2] During his editing of Black's lectures for the press, his friend and colleague John Robison was often severely critical of Black's lecture course.[3] Henry Brougham, however, remembering Black's lectures at a distance of half a century, delivered a striking encomium which recalled Black's low-pitched yet distinct voice, and above all his demonstrative experimental ability:

In one department of his lectures he exceeded any I have ever known, the neatness and unvarying success with which all the manipulations of his experiments were performed. His correct eyes and steady hand contributed to the one; his admirable precautions, foreseeing and providing for every emergency, secured the other. I have seen him pouring boiling water or acid from a vessel that had no spout into a tube, holding it at such a distance as made the stream's diameter small, and so vertical that not a drop was spilt. While he poured, he would mention the adaptation of the height to the diameter as a necessary condition of success [. . .] The long table on which the different processes had been carried on was as clean at the end of the lecture as it had been before the apparatus was planted upon it. Not a drop of liquid, not a grain of dust remained.[4]

This was certainly convincing testimony to Black's remarkable hand-eye coordination at a relatively advanced age, and Brougham also had a sense of how, late in his lecturing career, Black's self-presentation had already become historical in nature. Robison, increasingly dismayed with the difficulties of converting Black's lectures into a publishable form, found Black's theory of lime to be 'tedious beyond bearing, and the reader (of any information) cannot but see the keeping up of the great discovery till the very last.'[5] Brougham, by contrast, clearly recalling the same specific lectures, wished he 'could be once more allowed the privilege which I in those days enjoyed of being present while the first philosopher of his age was the historian of his own discoveries.'[6]

Since James Kendall's publication of materials from the chemistry students' short-lived Chemical Society (1785–6), and Carleton Perrin's careful and well-detailed studies of the views of Black and of some of his students during the years of the Lavoisian revolution in chemical science,[7] we have also gained an increasing sense of the ways in which students of chemistry and medicine at Edinburgh in this period were far from being a passive audience for their professors in the course of such epochal change. Rather, the rising generation of students were active, to a degree independent, and certainly non-negligible participants in the process of scientific change; indeed, in some respects they may even be described as the *avant garde* of the Chemical Revolution in Scotland.

This essay takes up the trail of research first blazed by James Kendall, to look more widely at the independent culture of student science in

Edinburgh during the period of Black's professorship. It firstly examines some statistical features of the medical student body from which the students of chemistry were drawn in the latter half of the eighteenth century, in order to derive some estimate of the role and significance of chemistry in its pedagogical setting of the medical school. It then proceeds to a partial reconstruction of student scientific culture by briefly surveying the independent institutions which the students originated, and the highly distinctive ethos which they formulated and maintained for the pursuit of science. From key examples drawn from the 1770s and 1780s, emerging from and otherwise immediately relatable to student scientific culture, it will also prove possible to construct an account of the chemistry of the period which makes it clear that it was not neatly enclosed within the confines of an agenda of purely physical chemistry – concerned only with the Lavoisian views of calcination, combustion, oxidation, acidification and the composition of water – but was equally concerned with the development of chemical physiology. This chemical physiology was phlogistically based, flourishing in the chemically controversial context of the time, and came to incorporate key concepts derived from Black's heat research as well as his phlogistonism. The aggressive and controversial aspects of such work, its enthusiastic embrace of issues of high theory and its disposition towards professorial science are, it will be suggested, considerably clarified once the ethos and conduct of student science are taken into account.

Students of Chemistry: Some Numbers

In the period of Black's teaching – from 1766 to 1795 – 3,765 students, counted as an aggregate of year-on-year attendance, signed the matriculation albums with a choice of chemistry among the courses they attended. This total is somewhat high if taken to refer to an absolute number of individual students, for some would attend the course more than once; hence the qualification 'year-on-year aggregate', which has greater accuracy with reference to totalling the sizes of class per annum. Any separate lists of students kept individually by professors usually indicated a higher number per annum, as much as 10%, than the matriculation albums, but such lists are not numerous enough to make any reliably thorough comparison. There were, however, numbers of students signing the albums but with specific subjects not yet chosen. If we allot such students according to the general proportion of matriculated chemists within medical school matriculants – c.30% per annum in this period – then Black probably taught upwards of 4,500 students over three decades, an average of 150 students per annum. This average was distributed over, and was a function of, the general rise in medical student numbers after the mid-century, an increase by a factor of c. 2.5. By the 1790s Black was often lecturing to a class of over 300.[8]

Chemical Subject Choice in the Student Medical Society Dissertations, 1750 to 1790

In addition to the papers delivered in the short-lived student Chemical Society, which existed from 1785 to 1786, students in the Medical Society regularly delivered papers categorised as 'Dissertations' or 'Questions'. Out of 534 such papers, 109 (around 20%) were on subject matter requiring chemical knowledge (i.e. preparations, uses and operations of drugs/medicines; analysis and/or discussion of e.g. airs, acids, mineral waters and their properties; crystal formation; respiration; digestion; fermentation; inflammability; heat; animal heat; etc.). Eighty-six of these latter (78%) were in the period of Black's tenure. The annual numbers of chemically-inflected dissertations in the Society, bobbing along around two per annum, rose significantly from 1780, most likely in response to the impetus provided by the new French chemistry.[9]

Chemical Subject Choice in Inaugural Dissertations for Degree of MD, 1750–1800

When choosing subjects for inaugural dissertations for MD graduation in the same period, only 44, around 4%, opted for specifically chemical topics. That 4%, however, included the notable and well-known theses of, among others, Joseph Black, William Cleghorn, Daniel Rutherford and Patrick Dugud Leslie.[10]

A number of points emerge from the foregoing statistical features of the medical student body. The study of chemistry both benefited from and contributed to the soaring reputation of the medical school, a reputation reflected in the overall steeply rising rate of expansion. This was particularly noticeable from 1775 onward, and international in character, as the medical school drew in increasing numbers of Irish, English, American and European students. Unlike Oxford and Cambridge, Edinburgh imposed no religious tests upon its students, and this resulted in a notable influx of English religious dissenters. The school was catering for two principal educational markets. Firstly there were intending physicians, who would either graduate at Edinburgh or continue on to Leiden, say, or Paris, for further instruction and graduation. Secondly, in addition to those who only attended chemistry, there was the cohort of students coming for two subjects only, namely anatomy and chemistry; these groups would have had the intention of pursuing not the learned, gentlemanly profession of physician, but the trade of apothecary, or surgeon-apothecary.[11]

Left to themselves, acting independently within their own institutions, students demonstrated a substantial and persistent interest in chemical and medico-chemical topics, but that level of preference was not exhibited in the subject choices for the inaugural dissertations of graduating physicians, which opted largely for specifically medical topics, particularly the discussion of individual diseases. Yet, although chemical inaugural dissertations were comparatively few, some of them undoubtedly attained the status of authentic and innovative contributions to experimental and theoretical research in chemistry, and this was a phenomenon observable before the advent of the undoubted stimulus to chemical research provided by Lavoisier. It would appear, then, at this limit, there was no firm and absolute distinction which reserved chemical research in the university as an exclusively professorial province. Joseph Black, indeed – himself a notable student-researcher in the 1750s – was happy to recommend his student William Cleghorn's theoretical dissertation on the matter of heat and its combination with gross matter to the attention of his chemistry class.

Student Societies

In the second half of the century there existed a number of student societies devoted to the pursuit of the sciences. The largest number of such societies were medical in character, and included the Virginia Club, consisting of members from the colony; the Hibernian Medical Society, composed of Irish students; and the Medical Society, after 1779 the Royal Medical Society, which had originated in the 1730s and became the most prestigious, prosperous, exclusive, elitist and long-lasting of the student societies, possessing its own premises, with a library, some laboratory facilities and an experimental committee. Its royal charter was obtained through the good offices of Dr Andrew Duncan, a former senior president of the Society, and the Scot and former Edinburgh professor Sir

John Pringle, then in the last years of his presidency of the Royal Society
of London. There was additionally in the 1760s the Newtonian Club, and
by the 1780s both the Natural History Society and the aforementioned
Chemical Society. The latter was clearly formed around the controversial
issues posed by Lavoisier's oxygen chemistry, but its members were not
the first set of students to devote themselves to critical and controversial
matters. It is rather the case that critical scientific debate and controversy
had become a noticeable characteristic of the conduct of student science
by the mid-1770s, a decade earlier than the advent of the Chemical
Society. It is possible to gain an at least preliminary sense of the complex
of attitudes which motivated the critical independence of students by
studying features of the annual presidential and other formal addresses
delivered to the Medical Society, together with informal remarks
occasionally found in student correspondence or diaries.

In 1778 the Society had received a commissioned portrait of Professor
William Cullen, painted by David Martin. The president, Caleb Parry,
speaking to the occasion, remarked,

> Allow me to add a few more words to those who are still to remain.
> You have often in your memory, and have now constantly before
> your eyes [gesturing, one imagines, to the portrait of Cullen],
> him, whose character you have long and justly reverenced. May his
> presence fire you with zeal to imitate so bright a pattern. Imagine
> he inspects your conduct, and hears your debates; and act up to
> that example which he, and his joint teachers, have manifested in
> their own persons.[12]

A comparable, private, and more succinct reference to Cullen as tutelary
patriarch of the Medical School was made by Sylas Neville in 1775, after
the ceremony for the laying of the foundation stone for the Society's new
Hall: 'Was present at laying the foundation stone of a Hall for the Med.
Soc. It was laid by our father of medicine, Dr. Cullen [. . .]'[13] Two years
earlier, senior president Andrew Duncan had strenuously emphasised the
potential rewards of Edinburgh's medical education: 'By united endeavours, and by noble emulation, great and rapid progress in the knowledge
of this science may with confidence be expected; and your most sanguine
expectations in resorting to this school of medicine be answered.'[14] Caleb
Parry, elaborating more generally upon the above invocation of Cullen,

did not restrain the anticipations of his audience: 'Possessed, as you are, of the richest means of acquiring excellence, if you exert your own best endeavours, you cannot fail of ensuring that success which your professors have obtained, and may become, like them, the DICTATORS of the MEDICAL WORLD' (capitalisation in original).[15]

Such examples convey an impression of self-confident entitlement, and of anticipation of elite medical authority, to be attained by what Duncan called 'emulation', a self-conscious, competitive striving to equal or surpass the achievements of one's pattern-setting seniors. These seniors, as models to emulate, a Cullen or a Black, were also respected, even revered as such, within a homosocial sense of social and institutional structure which wore its patriarchy as it were on its sleeve, and with evident pride. This does not yet adumbrate, however, the equally strong sense evident in the Medical Society of the critical independence claimed by the students. Gilbert Blane, speaking on the occasion of the laying of the new Hall's foundation stone, addressed his fellow students as follows:

> I will venture to appeal to every one's experience, if, in the glow of social debate, he is not conscious of a vigorous exertion of mind, of an energy of thought unknown in the solitary hour. To discover truth, to detect falsehood, to develop the seeds of genius, and to emancipate the mind from the fetters of authority and prejudice, were the grand objects of this institution.[16]

Here, Blane picked out firstly the distinctive, cognitive sociability of the Society, the 'exertion' and 'energy' released in the process of oral debate, with its implied capacity both to liberate the enquiring mind from existing bias and to overcome the limitations of currently authoritative doctrine.

The key feature of debate in the Society is intelligible in terms of its being held in place by the way in which the routines of the Medical Society mimicked the course of exercises prescribed for taking the degree of MD, which had pronounced oral components, and which could require formally disputatious verbal skills from students required to defend whatever doctrines and arguments their work had espoused. A term which usefully covers this increasingly characteristic feature of the conduct of student science is 'agonistic', capturing as it does the sense of

combative, critical struggle involved in intellectual debate. Enlarging on
the ethos which the Society was held to promote, Blane insisted:

> Our predecessors perceived that it was not merely in the frigid
> plodding on books, nor the doctrines and precepts of age and
> authority, nor the little detail of an empirical practice, that could
> inspire that taste and spirit, and give that manly turn to our inquiries,
> which alone can render study agreeable, vigorous and successful.
> They perceived that it was in Society alone, by the mutual commu-
> nication of the lights of reason and knowledge, that the intellectual
> as well as the moral powers of man are exalted and perfected.[17]

Here the attainment of independent masculine selfhood, 'taste', 'spirit',
the 'manly turn', was recruited as the gain of the Society's oral, agonistic
sociability, contrasted with the sterile, monkish and scholastic procedures of former days.

It is worth pondering, however, just what was meant by Blane's ascrip-
tion of these foundational cognitive virtues to 'our predecessors', some
of whom were of course present at the occasion of Blane's address. The
contemporary professorial generation, represented by the likes of Cullen,
Black, Alexander Monro *secundus*, and earlier by Robert Whytt, were
notable in intellectual and pedagogic terms for their radical and innovative
character. Collectively, their teaching had changed the face and function
of chemistry, re-gearing it as a modern, 'philosophical' science rather than
a pharmaceutical art, and in medicine deserting the Boerhaavian ortho-
doxies of the three decades between 1726 and 1756 to produce a new and
productive focus upon, and investigation of, the nervous system. These
contemporary models of emulation, the professors, did not frigidly plod
on at their books, and they successfully and controversially had resisted
the precepts of age and authority current in their younger days to produce
their own, powerful new systems of chemistry and medicine. Blane was
attempting to frame the expansive and agonistic virtues as the *leitmotif*
of the Society's history, from its origins in the 1730s onward, to the effect
that the current generation of students were now its inheritors. Truly and
thoroughly to emulate their predecessors, current students might thereby
find themselves in critical opposition to those predecessors.

Were one a professor in the 1770s, a degree of ambivalence towards
the students' enthusiastic agonism might have been understandable,

and in the event, entirely reasonable. The Medical Society's members did conduct debates in the 1780s, persistently returning to fevers and their treatment, and to the medical effects of opium. Such issues were subsumed, however, by the polarisation of the students into two warring parties, one of which enthusiastically took up the doctrines of John Brown, a former pupil of Cullen's, whose medical system produced a drastically simplified physiology, nosology, diagnostics and treatment regimen. The other party remained loyal to the doctrine and system of Cullen. These Brunonian debates were a complex historical phenomenon which cannot be further pursued here, other than to remark on the way in which they forcibly illustrated the critical, agonistic ethos prevalent in medical student circles, indeed offered the minatory spectacle of what happened when the spirited intellectual antagonisms of the students utterly abandoned the manners of polite and gentlemanly scholarly behaviour.[18] From the viewpoint of history of chemistry in Edinburgh in the 1780s, two further points may be made. The agonistic liberty of the students virtually guaranteed that when a new, powerful and controversial form of chemistry arrived from France, it would be embraced with keen and institutionalised interest by some students, as it was in the Chemical Society. This interest and the arguments which accompanied it were, however, in comparison with the ferocity of the Brunonian debates, a relatively sedate sideshow in the spectacular theatre of Edinburgh scientific controversialism in the 1780s.

The emergence of publicly critical attitudes towards professorial science considerably pre-dates these controversies of the 1780s. In 1778, Irishman Patrick Dugud Leslie published his *Philosophical inquiry into the cause of animal heat*, a work which took up and extensively elaborated the topic of his inaugural dissertation of 1775, *De calore animalium causa*. In the *Philosophical inquiry*, he devoted an early chapter to a critical survey of earlier and contemporary accounts of animal heat, developing first a critical rejection of mechanical (i.e. motion-based) accounts. He then advanced to a critique of the views of Edinburgh's two chief chemists, his teachers Professors Cullen and Black. Cullen's vitalist account ascribed animal heat to 'some circumstance in the vital principle of animals, which is in common to those of the same class'.[19] Nevertheless, Leslie continued, 'while we admire the singular ingenuity, which stamps every part of the *Cullenian* doctrine, we must be permitted to consider it, in this particular, as founded on a more specious than solid basis. What grounds have we to imagine the principle of life different in different animals [. . .] Upon the whole, from these views of the many objections that tend to overturn Dr. *Cullen's* theory of animal heat, we do not hesitate to account it a mere hypothesis, and entirely abandon it.'[20]

Black's theory (possibly prompted by Virginian student James McLurg's 1770 dissertation, *De calore*), received altogether more approval, as 'perhaps the most ingenious and best supported theory ever proposed upon the subject of animal heat'.[21] Black had observed the proportionality between rate of respiration and degree of animal heat, and concluded that it was generated in the lungs by air's action upon phlogiston, and then diffused by the circulatory system: respiration as slow combustion. Despite the apparently persuasive evidence for Black's views, Leslie commenced a series of animadversions upon them.

> These arguments may, perhaps, on a superficial view of the question, appear conclusive; but a sound reasoner, who shall coolly and impartially weigh every circumstance, will, I am confident, allow that they only afford a very ambiguous and imperfect evidence of the doctrine, which they are meant to establish, and the subsequent animadversions on Dr. *Black's* theory at large will, it is hoped, suffice to show that it is not only founded on dubious and controversial principles, but that it is, in every point of light, clogged with insurmountable difficulties.[22]

Even the best of professorial efforts, it appeared, fell far short of inducing conviction. Leslie did not object to Black's phlogistonism, but in particular did not like Black's site-specific location of heat generation in the lungs. He welcomed instead Joseph Priestley's new theory of respiration (dephlogisticated air (oxygen) inhaled, surplus phlogiston derived from nutrition circulated to lungs, and exhaled), and eventually commended a theory he ascribed to Andrew Duncan, luminary of the Medical Society.[23] This equally phlogistic theory had animal heat evolved in the whole course of circulation, as freed phlogiston was released within the body by the capillary action of the blood vessels.

Leslie went on to a speculative elaboration of phlogiston within nature as a whole, equating it with the aether of Isaac Newton, and thereby identifying it also with current material theories of light, heat and electricity, a phlogistic philosophy of nature, and, although broader, not dissimilar to the phlogistic views adopted in John Robison's Edinburgh lectures on natural philosophy.[24] Phlogiston was, by this stage in its Edinburgh career, undergoing an expansive usage, advancing from its chemical origin effectively to colonise both physiology and natural philosophy.

This advance continued the following year with the publication of a work by another Irish chemical physiologist, Adair Crawford, a Glasgow graduate. Crawford developed a more complex, in its own terms highly coherent theory of animal heat. It separated heat and phlogiston, in contrast to Leslie's identification of them, and used his Glasgow teacher William Irvine's notion of 'absolute heats' (calculated on specific heat capacities, and incorporating latency as a modification of capacity), itself a development of Black's latent heat, to propose a cyclical series of heat-phlogiston exchanges during circulation, the heat originating as 'fixed' or latent heat in the air inhaled into the lungs. Unlike Leslie, Crawford used his own experimental work in his calorimetrically significant measurement of animal heat, and he would go on to become a principal defender of phlogistic chemistry in Britain, along with Richard Kirwan and Joseph Priestley.[25]

The relevance of Crawford's work to the Edinburgh students is both striking and immediate. Crawford prefaced his work with a brief historical account of it, describing the original experiments as having been carried out in Glasgow in the summer of 1777, these experiments then being made known to Glasgow professors Thomas Reid, Patrick Wilson and William Irvine. The experiments were then made known to the Edinburgh professors in the autumn of 1777, and were relayed specifically to the students in the ensuing academic session by a lecture to the Medical Society in the following winter.[26] Crawford's research thus had a notable pre-publication itinerary which included its exposure to the students of the Medical Society, as interested parties, and part of the first-phase audience for the research.

The works of Leslie and Crawford, together with Joseph Black's concurrent advocacy of the negative-weight hypothesis for phlogiston, constituted the high point of Edinburgh's phlogistic renaissance; but over the next five years this enthusiasm would fade, Black forsaking negative weight, while some of his students started to adopt Lavoisier's chemistry by 1784–5.[27] Given the intellectually bellicose attitudes present among the students at just this time, Black may also have judged it impolitic to engage in direct debate with advocates of the new French chemistry. In the Royal Society of Edinburgh the new chemistry was debated in 1788, James Hall speaking for the new chemistry, James Hutton defending phlogistic chemistry. This debate had a discernible impact upon Robert Kerr, who recalled that Hall managed to 'shake the phlogistic faith of many, amongst whom I was'.[28] A defining moment for chemical pedagogy in Edinburgh soon followed. In September and October of 1790, Kerr swiftly translated Lavoisier's new textbook of 1789, the *Traité Elémentaire*, the haste required because 'it was judged necessary by the Publisher that the Translation should be ready by the commencement of the University Session at the end of October'.[29] Although Hutton continued to defend the phlogistic faith in the 1790s, and although Robison bemoaned the *a priori* synthetic form and new vocabulary which he perceived as characterising French chemistry, in practical and pedagogic terms the issues were mostly decided, for the students were reading and speaking the new language of chemistry.[30]

The ethos of critically independent, at times anti-professorial agonistic liberty, and the persistent engagement of students with major theoretical and experimental matters in chemistry and chemical physiology, were, it has been argued, prevalent and distinctive features of Edinburgh chemical life, running through both its phlogistic and Lavoisian phases, and having authentic, discernible and major impact upon the development of chemical science throughout this period. When acknowledging the positive reception of the new chemistry in Edinburgh, Joseph Black wrote to Lavoisier that 'although the power of habit may prevent many of the older chemists from approveing [sic] of your Ideas, the younger ones will not be so influenced by the same power; they will universally range themselves on your side of which we have experience in this university where the students enjoy the most perfect liberty in chuseing [sic] their philosophical opinions'.[31]

The previous two decades had given the Edinburgh professoriate ample experience of the truth of this assertion. Indeed, given the intensity of debates in chemistry and medicine and the rapid pace of change, it might be concluded that Black had discreetly understated, rather than exaggerated, the case.

Notes and References

1 Bell, Whitfield J., Jnr., 'Thomas Parke's Student Life in England and Scotland, 1771–1773', *The Pennsylvania Magazine of History and Biography* 75 (1951), p. 249.

2 Cozens-Hardy, Basil (ed.), *The Diary of Sylas Neville, 1767–1788* (Oxford: Oxford University Press, 1950), pp. 198, 216.

3 For an account of Robison's fraught editorship of Black's lectures, see Christie, J.R.R., 'John Robison and Joseph Black', in A.D.C. Simpson (ed.), *Joseph Black, 1728–1799: A Commemorative Symposium* (Edinburgh: HMSO, 1982), pp. 47–52.

4 Brougham, Lord Henry, *Lives of Men of Letters and Science who flourished in the age of George III* (London: Knight & Co., 1845), pp. 346–7.

5 Letter, John Robison to James Watt, Edinburgh 23 July 1800, in Robinson, Eric and Douglas McKie (eds), *Partners in Science: Letters of James Watt and Joseph Black* (London: Constable, 1970), p. 344.

6 Brougham, *Lives of Men of Letters and Science*, p. 348.

7 Kendall, James, 'The First Chemical Society, the First Chemical Journal, and the Chemical Revolution', *Proceedings of the Royal Society of Edinburgh* 63A (1952), pp. 346–58, 385–400; Perrin, C.E., 'A Reluctant Catalyst: Joseph Black and the Edinburgh Reception of Lavoisier's Chemistry', *Ambix* 29 (1982–3), pp. 141–76; Perrin, C.E., 'Joseph Black and the Absolute Levity of Phlogiston', *Annals of Science* 40 (1983), pp 109–37.

8 Figures compiled from Matriculation Albums 1750–1800, Edinburgh University Library, Manuscripts and Rare Books (ref. Da.).

9 Figures compiled from 'Dissertations read to the Royal Medical Society, Edinburgh', vols 1–42 (1750–1800), British Online Archives, Royal Medical Society of Edinburgh.

10 Black, Joseph, *Dissertatio medica inauguralis, de humore acido a cibis orto* (Edinburgh, 1754); Rutherford, Daniel, *Dissertatio inauguralis, de aere fixo dicto, aut mephitco* (Edinburgh, 1772); Leslie, Patrick Dugud, *Dissertatio medica inauguralis, de caloris animalium causa* (Edinburgh, 1775); Cleghorn, William, *Dissertatio physica disputatio, theoriam ignis complectens* (Edinburgh, 1779).

11 Rosner, Lisa, *Medical Education in the Age of Improvement: Edinburgh Students and Apprentices, 1760–1826* (Edinburgh: Edinburgh University Press, 1991) provides further detailed analysis of the composition of the Medical Student body. For the topics discussed here, see particularly ch. 6, pp. 104–34.

12 Parry, Caleb, *Closing Presidential Address*, Medical Society (Edinburgh, 1778), p. 5. The portrait of Cullen to which Parry referred was by David Martin. It was joined by a portrait of Joseph Black, also by Martin, and they hung together as a pair on the same wall in the Hall of the old premises of the Royal Medical Society. Both are now in the Scottish National Portrait

Gallery. The Black portrait adorns the front cover of this volume.

13 Cozens-Hardy, *The Diary of Sylas Neville.*

14 Duncan, Andrew, *Closing Presidential Address,* Medical Society (Edinburgh, 1773), p. 6.

15 Parry, *Closing Presidential Address.*

16 Blane, Gilbert, *Address to the Medical Society of Students at Edinburgh, upon the laying of the Foundation of their Hall,* Medical Society (Edinburgh, 21 April 1775).

17 Ibid.

18 For Edinburgh Brunonianism, see Barfoot, Michael, 'Brunonianism under the bed: an alternative to university medicine in Edinburgh in the 1780s', in W.F. Bynum and Roy Porter (eds), *Brunonianism in Britain and Europe, Medical History Supplement No. 8* (London: Wellcome Institute for the History of Medicine, 1988), pp. 22–45; and Risse, Guenter B., 'The Royal Medical Society versus Campbell Denovan', in *New Medical Challenges in the Scottish Enlightenment* (Amsterdam and New York: Rodopi, 2005), pp. 105–132.

19 Leslie, Patrick Dugud, *A philosophical inquiry into the cause of animal heat: with incidental observations on several physiological and chymical questions* (Edinburgh: Gordon & Eliot, 1778), p. 69.

20 Ibid., p. 72.

21 Ibid., p. 75.

22 Ibid., p. 77.

23 Ibid., pp. 92–3, 159–60.

24 See Wilson, David B., *Seeking Nature's Logic: Natural Philosophy in the Scottish Enlightenment* (University Park, PA: Pennsylvania State University Press, 2009), pp. 226–7.

25 Crawford, Adair, *Experiments and observations on animal heat, and the inflammation of combustible bodies. Being an attempt to resolve these phaenomena into a general law of nature,* 2nd edn (London: Murray, 1788 [1779]), passim.

26 'Advertisement', in ibid., p. i.

27 See Perrin, 'A Reluctant Catalyst'.

28 Letter, Kerr to Lavoisier, 21 January 1791, cited in McKie, Douglas, 'Antoine Laurent Lavoisier, F.R.S. 1793–1794', *Notes and Records of the Royal Society of London* 7 (1949), p 13.

29 'Advertisement', in Kerr, Robert (trans.), *Elements of Chemistry in a new and systematic order, containing all the modern discoveries . . . By Mr. Lavoisier* (Edinburgh: William Creech, 1790), p. vi.

30 Hutton, James, *A Dissertation upon the Philosophy of Light, Heat and Fire* (Edinburgh: Cadell, 1794); Letter, Robison to Watt, 23 July 1800, in Robinson and McKie, *Partners in Science*, p. 345.

31 Anderson, Robert G.W. and Jean Jones (eds), *The Correspondence of Joseph Black* (Farnham: Ashgate, 2012), vol. 2, p. 1101.

SIX

Useful Pictures:
Joseph Black and the Graphic Culture of Experimentation

MATTHEW DANIEL EDDY

Introduction

In the history of science Joseph Black is best known for his isolation of fixed air (carbon dioxide) in the 1750s. But in addition to his experimental research, he taught chemistry in the University of Edinburgh's Medical School for the last four decades of the eighteenth century.[1] While his students were certainly impressed with their professor's chemical expertise, they also regarded him as a skilled teacher and a gifted communicator. Crucial to his teaching was a carefully curated assemblage of diagrams and figures that depicted theoretical and practical aspects of chemical affinity. They occur in most student notebooks taken in his lectures and, though studies on Black's chemistry cite them in reference to his experimental research, they are seldom treated independently as visual objects of inquiry that were designed primarily for teaching students. Building on recent research that underscores the pedagogical and visual facets of Black's diagrams,[2] this essay further explores how Black's visualisations were relatively simple pictures to which he attached chemical meanings that were easy for his students to understand.

Throughout the four decades in which he taught at Edinburgh, Black employed simple visual structures to represent how substances were attracted to each other in compounds. During the eighteenth century most chemists held that substances were bound together by an invisible force of attraction called affinity. They did not, however, have a unit to measure the force of attraction. But, based on painstaking experimentation, they knew that most substances had a stronger or weaker attraction to other substances. Thus, in simple reactions, they could predict which substance would unite with another. This kind of attraction was called 'single elective affinity'. For more complex reactions, they used the term 'double elective affinity' to try and explain how the competing attractions of different substances resulted in the final products.

Black used two kinds of pictures to visualise various aspects of single and double attractions. First, he employed figures to depict instruments and experimental tableaux. Such figures were mimetic pictures, in that they sought to represent objects that could easily be seen by the human eye. Secondly, he used diagrams to represent scientific concepts like heat and force that could not be observed directly by the human eye. These diagrams were schematic in that they reduced unobservables to simple lines and patterns that could be easily remembered and quickly drawn. To visualise single elective affinity he used the square, or, more specifically, he used a square-shaped affinity table. For double elective affinity, he used a circlet diagram to represent a basic reaction and a chiastic diagram to represent the ratios of attraction between substances. When used together, the three diagrams – the table, the circlet and the chiasm – collectively visualised the theoretical underpinning of his chemistry and, hence, functioned as a visual system (Plate 2).

When Black's students attended his lectures, they had to learn how to use his figures and diagrams as pictures that could be read in a number of ways. This means that these pedagogical images consisted not only of a visual form, but also of a set of practices used to learn, make and understand that form. They were visualisations that students were taught to think with, things that shaped what they knew and what they thought they knew. It is this active view of an image that is used in the following sections to unpack Black's figures and diagrams, especially the notion that they were visual forms which gained meaning and value as they moved through time and space in the notes kept by professors and students. The intention is to gain deeper understanding of the role they played as learning tools and informatic devices that made chemistry easier for students to understand.

A Graphic Continuum

The most influential visualisation used to teach chemistry during the eighteenth century was the affinity table proposed by the French academician Étienne François Geoffroy in 1718. Even though it was a 'table', its rows and columns read more like a diagram. It consisted of lists of substances and it was reproduced and modified throughout the century.[3] This kind of visualisation was used in an age when both tables and polygons were considered images. In Britain this view of representation was expressed in the works of the philosopher and physician John

Locke, whose views on the cognitive efficacy of graphic culture were taken up by Scottish intellectuals during the mid eighteenth century. Consequently, most works written by Scots on pedagogy promoted an educational psychology that advocated the use of images made from words and simple lines, thereby transforming simple schemata or even lists into crucial learning tools that also served as information management devices. This view of pedagogical pictures treated them as knowledge-making artefacts and, as such, it resonates with current research in anthropology, art history, psychology and the history of science.[4]

Many of Black's figures and diagrams were similar to those used in many early modern institutions which taught medicine and natural philosophy.[5] But unlike the three-dimensional figures of body parts and experimental apparatus depicted in reference books designed for popular consumption, pictures used for medical instruction in Scotland tended to be more schematic because they were often being used alongside the anatomical specimens and instruments that they were meant to represent. Black and his Edinburgh colleagues were heirs to the graphic techniques of this tradition.

Black's pictures existed in a temporal continuum of graphic iteration and innovation. As intimated above, some extended contemporary graphic traditions in medical chemistry and natural philosophy. Others, such as the affinity table, were adaptations of forms taken from the longstanding mnemotechnic tradition. Additionally, he changed them over time as needed throughout the course of his career. In this respect – that is, in relation to their usage – his pictures were, firstly, part of a general tradition of schematic representation and, secondly, part of the particular chronology of each chemist who used or modified them. Tracing these kinds of modifications or changes in the world of Enlightenment chemistry is generally difficult because it is a challenge to find extant sources that are appropriate.

Fortunately, Edinburgh's chemistry students left behind many notebooks. These show that the concept of using chemical visualisations like diagrams and figures had already been firmly established by William Cullen, Black's teacher and mentor.[6] They also reveal that, though the basic chiastic, circular and tabular forms of Black's diagrams remained relatively consistent over his career, he modified them in a number of ways. Additionally, his affinity table and its symbols – like most affinity tables used by Enlightenment chemistry teachers – also built on a visual tradition introduced by Geoffroy in 1718.

Black was very aware of the foregoing temporal continuum of visualisation, and commented on how the arrangement and content of his affinity table either extended or departed from graphic forms of representation used in the distant and recent past. This awareness is perhaps most clearly evinced in the sections of his lectures that explained the meaning of his chemical symbols. When the Swedish chemist Torbern Bergman definitively isolated a substance called calcareous earth during the 1770s, for instance, Black appropriated Bergman's symbol for it. In the lectures he gave over the next two decades, he took care to tell his students how and why Bergman had constructed the symbol from previously existing 'marks'.[7]

Black also redesigned some of his chemical symbols so that they could be more easily remembered by his students. Notably, he felt it was important to explain the historical context of his decision. The best example of his practice can be seen in his redesign of the symbol used for general alkalis. Black explained the change to his students in the following manner: 'For Alkalis in gen[eral] I use a Circle with a semi Circle added to one side. The Mark used in Geoffroy's Table is very diff[erent], but it is like the mark for Vitriol [in that] it is not easy to remember it, & thus is more simple and distinct.'[8]

Black's diagrams also extended local Scottish traditions of representation. His circlets extended the kind of geometric circles used in university natural philosophy textbooks, and his chiasm was a redeployment of a graphic calculation device used by Reformed schoolchildren. But his iterations of such images were not straightforward replications, and he made alterations to them throughout his career so that he could visualise new developments in chemistry. Thus, as his career progressed and he learned more about chemistry and teaching, he transformed the numeric ratios of his chiasm into algebraic equations and he inserted additional substances into his circlets.[9]

Black's figures and diagrams were more like snapshots, idealised moments that were frozen in time. In this respect they were atemporal and this meant that Black had to communicate the temporal instructions verbally. Even when students attended lectures and actively sought to connect the meaning of Black's pictures to his hands-on experimental practices, they still had to grapple with the fact that he had often done part of the experiment prior to their arrival in the classroom.[10] This means that Black's figural depictions of experimental tableaux were in many ways an attempt to mitigate this temporal conundrum (Plate 3).[11] But the important point to note here is that the number and order of steps required to conduct a successful experiment were explained in the *verbal* instructions, and not strictly in Black's *visual* depictions.

Since Black's pictures were *prima facie* atemporal depictions, it was rather difficult to discern their temporal meanings when students first encountered them. It was only when students watched Black conduct and explain his experiments in the classroom that they learned to attach elements of temporality to them. When read one way, for example, the chiasm could be interpreted to depict substances that were together at the *start* of a reaction (Plate 2a). When read another way, the chiasm could be interpreted to represent the substances that were together at the *end* of the reaction. This multistable aspect of Black's pictures made them versatile, but it also meant that they were difficult to interpret without having attended his lectures or, at the very least, without having access to a very thorough set of notes that were taken in his class.[12]

The temporal complexity of Black's pictures can also be seen in the alphabetical headings that he used to label his figures. The narrative instructions that corresponded to such headings were effectively step-by-step procedures which governed the order of the actions that needed to be taken in a set of multi-stage experiments.[13] As such, the figures and their headings corresponded to lists of experimental events or episodes, all of which added up to a final product that could be explained through the affinity model. The best way to understand the temporal dimensions of Black's figures was, again, to watch him perform the experiment in person. Barring this form of multi-sensual learning, students who missed the lecture had to resort to the oral or written accounts of other students.

Attaching Time to Figures

Black made his figures visually simple so that students could easily associate them with what they had seen in lectures. His figures came in two schematic varieties. The first were small pictograms. The objects of these pictograms were usually instruments and students tended to insert them between the words or sentences in their notebooks. The second were larger, more developed drawings that some students shaded with cross-hatching or, in a few cases, watercolours.[14] Like the pictograms, the objects of these drawings were instruments; however, they also included additional objects or features that helped determine the placement of the apparatus during an experiment. As shown in Plate 2b, the greater detail of these drawings made it possible to see how instruments and supporting apparatus (like connecting tubes) were set up during the experiment. In some cases, the shading or contour lines revealed modifications or singularities to the specific instruments that Black was using in his classroom experiments.

It is clear that Black gave his students some sort of standard visual depiction of important instruments or experimental setups. The distribution of figures in this manner, either as a handout or as a poster hung at the front of the room, was practised by other members of the Edinburgh medical school. Black's mentor William Cullen, for instance, hung affinity tables at the front of his classroom.[15] Black most likely presented his figures as a poster or as a handout, because the same ones appear in different sets of notes and many student drawings feature alphabetical or numerical headings that, though clearly relevant to the experiment under consideration, are not verbally explained in the notes. This strongly suggests that students simply copied the figures but then did not have time to record the meaning of the headings. It also suggests that graphic simplicity facilitated their preservation.

The use of schematic figures in this manner points to a fundamental interpretive question that is relevant to Black and most of the Enlightenment teachers who used graphic techniques to represent natural knowledge: how was the design of the figure influenced by its intended use? In Black's case, the intended users were the hundreds of medical students who attended his chemistry lectures during his tenure at Edinburgh. This means that, rather than being the 'graphic gropings' found in many laboratory notebooks, they were carefully designed to be simple visualisations that could be easily drawn and remembered.[16]

When it came to understanding the affinity concept, Black's figures also had a number of conceptual advantages. At one level they connected theory with practice because they helped to illustrate what a student could do materially in a lab with the affinity concept as represented by his chiastic and circlet diagrams, and by his affinity table. At another level they functioned as information management tools, serving as mnemonic encapsulations that helped students make sense of the step-by-step experimental instructions given during the lectures.[17] The figures also allowed students more easily to associate the invisible attractions of substances with the visible properties evinced through observation and experimentation. The main way that Black facilitated this act of association was by allowing his students to experience the smells, bangs and colours of chemical reactions.[18] While students preserved these sensations verbally in their notes and imaginatively in their minds, Black's figures of instruments such as Florence flasks, tubes, furnaces and the like functioned as schematic memory aids for what they had seen, heard and smelt.[19] In this sense, his figures were a visual reminder of chemical attractions that his students had already experienced for themselves.

It was difficult for students to interpret Black's figures without the aid of marks or symbols associated with the passage of time.[20] The experiment depicted in Plate 3, for example, would be hard to discern were it not for the alphabetical headings that corresponded to the verbal experimental instructions written in a student's notes. Likewise, the figures were usually presented either partially or totally in one dimension, thereby removing unnecessary spatial distractions. The vessel on the right side of Plate 3, for instance, is a good illustration of this kind of schematic flatness. The figure, which is taken from a copied set of notes that Black had transcribed from his own lecture notes, is a particularly good specimen. Many student copies are even less detailed. The necessity of attending lectures to take notes that explained the figures dovetailed with an important pedagogical principle of deictic learning: the material act of inscribing further engrained the chemical products and processes that Black attached to his figures.

At the simplest level, the figures encapsulated various intervals of time that elapsed in chemical reactions. Some experiments – those with acids, for example – happened quickly, sometimes in seconds. Others, such as distillation and crystallisation, could take longer periods of time – sometimes days, or even weeks.[21] To overcome these temporal limitations, Black used substances that had been prepared in advance and summarised the steps through which the final products had originally been obtained. Regardless of how long his experimental demonstration took to prepare outside the classroom, or to perform in front of students, the figures represented chemical processes at a glance as frozen points in time.

Like so many chemical visualisations of the Enlightenment, Black's figures erased the time of all the failed experiments that had contributed to the isolation of the substances which they often featured.[22] Even though his figures were largely atemporal, it took a good amount of time for his students to learn how to use them when they were not read alongside a set of lecture notes. A case in point can be seen in the rough notes taken by Charles Blagden in Black's 1766 lectures.[23] Blagden would go on to be the Secretary of the Royal Society of London and one of Black's most influential students. The first time that Blagden encountered Black's circlet diagrams, however, he struggled to reproduce them in his notebook. In the end he learned to use them by shaping them as squares – an act that took some time to work out and then to copy into his notes (Plate 4).[24]

The general point to take from Blagden's unfinished circlets, or even from the many complete versions that appear in other student notebooks, is that, when the figures are treated as information management devices, it can be seen that they required an assemblage of skills and routines to use them and to make sense of them. In addition to their basic geometric structure, students had to learn, for example, the meaning of the chemical symbols and how to follow the flow of information. As Blagden's case illustrates, since there were no printed versions of the diagrams, students also had to learn how to draw them in their notebooks.

Designing Space in a Table

As evinced in the tables of Black's contemporaries, most notably in the lectures of Gabriel François Venel in France and Richard Watson in England, the traditional alignment used to organise affinity tables was a vertical column of substances listed from top to bottom (Plate 5).[25] The symbol of the main substance was placed at the top and the substances attracted to it were listed below it. The order of the list corresponded to the strength of the attraction, with the strongest at the top and the weakest at the bottom.

Black turned the traditional affinity table format on its side, transforming the columns into rows and vice versa. His table, therefore, was based on rows that were read from left to right. The main substances ran down the left side and their affinities ran horizontally, thereby positioning the strongest attractions on the left and the weakest on the right (Plate 2c). Such a structure facilitated the general left-to-right reading pattern present in sentences and the dichotomous chemical tables used by Black and other contemporary chemists in the Scottish university system and elsewhere.[26] While Black's table was unique in several ways, the notion of using rows to order substances was already known in Scotland during the mid eighteenth century. A similar alignment occurs, for example, in the affinity table of William Lewis's 1753 *Dispensatory* (Plate 6).[27] Black cites it in his 1760s lectures, and Lewis's graphic layout probably influenced his decision to make a table that read from left to right.[28]

The internal organisation of most affinity tables was based on groupings of substances that shared some sort of similarity. Geoffroy initiated this tradition by grouping salts (acids and alkalis) on the left side and metals on the right side, with water being added as a final column. Yet the meaning of his grouping, which was iterated by many mid to late eighteenth-century teachers (such as Venel), would not have been immediately apparent to students, especially since at first glance it might have seemed to them that Geoffroy had misplaced some substances. For example, it is likely that a new chemistry student would struggle to intuitively understand why Geoffroy placed *soufre mineral*, an inflammable substance, in the metallic grouping. Geoffroy's table also contained other potential classificatory confusions. The placement of the substances of course made sense when explained in lectures, but the disconnection between the conceptual order and the visual order remained a graphic problem that needed to be solved.[29]

Like Geoffroy and Venel, Black's table began with salts (acids and alkalis) and moved on to metals (Plate 2c). Since Black had turned the table on its side, the former occurred at the top and the latter appeared in the middle. Next came 'mild substances', and then a last group of substances formed by special heating factors. The order of Black's table shows that he followed the general practice of keeping salts and metals together in groups. But the crucial difference that distinguished his table from others was that it featured two graphic innovations that effectively eliminated the conceptual and visual disconnections outlined above.

First, Black broke with the practice of using a singular grid to house all the substances. Instead, he made each grouping a separate block of information. The end result was that each grouping became a visually distinct unit of information that students called a 'divisio' or 'part' in their notes. Second, Black included headings that explained the relationship between the substances contained in each block. For instance, above the first grouping, he explained its contents with the following head: 'Containing ye Relation of Alkalis and Alkaline substances to Acids &

Substances of an Acid Nature' (Plate 2c).[30] The use of such headings was a crucial pedagogical innovation because it clearly laid out the organising principle that he had used to group the affinity reactions in his table, thereby making it easier to use and understand.

Black's students could use and replicate his table in several ways. When all the groupings were viewed as one visual structure (a module, so to speak), it offered a systematic overview of the single elective attractions that played a central role in late eighteenth-century chemical experimentation conducted in medical, industrial and academic settings. When one 'part' (or 'division'), which was really a 'microtable', was read on its own, it offered a succinct overview of how affinity operated in relation to an analogous set of substances reacting in a similar way to create different compounds. Most students attempted to keep the microtables together on the same page. Paul Panton's table in Plate 2c is a good example of this practice. Other students – Thomas Cochrane, for instance – copied the microtables onto separate pieces of paper, thereby isolating one form of attraction in a way that made it easy to read and remember.[31]

Black's use of separate headings for each microtable appealed to students as well, especially when it came to the personalised nature of note-taking. Instead of copying the headings in his table verbatim, students sometimes customised the wording so that it suited their learning needs or so that it could fit into the layouts of their notebooks. This can be seen by comparing the wording of the headings used by Cochrane and Blagden during the 1760s for the second microtable. Even though the contents of both tables were the same, the headings were slightly different. Cochrane's heading reads: 'The diff[erent] Attractions of Alkalis for Acids'; while Blagden's heading reads: 'Attractions of Acid substances for [general alkali symbol] and & [metallic substance symbol]'.[32] Additionally, whereas Cochrane's heading used no chemical symbols, Blagden's heading, as intimated inside the brackets inserted into the foregoing quotation, used the chemical symbols for general alkalis and metallic substances, eliminating the need for him to write out their names.

The simplification offered by microtables and their associated headings was that they structured the page in ways that saved students from having to trawl through all the columns of the rapidly expanding affinity tables being made from the 1760s forward. Black's segmentation of the affinity table into microtables effectively created detachable visual units that could be inscribed as needed in different parts of a student's notebook, or which could be copied onto a piece of loose-leaf paper and then used alongside a specific section of the notes. This aspect of the microtables had an added advantage in that it allowed Black to visually refocus his students' attention on a set of related reactions. This advantage made it easier for students to copy the visualisation into the specific section of their notes that addressed the relevant reactions. In other words, Black's detachable microtables offered a friendlier visual format to students, mainly because they were easier to use.

Black was able to keep his microtables relatively compact because he limited the number of substances that appeared on them. Here we can see him acting as a thoughtful teacher who conscientiously mediated chemical information so that his students would not feel overwhelmed with the flood of new substances that were being discovered during the second half of the eighteenth century. Some professors tried to keep pace with these discoveries by adding many new columns to their affinity tables.[33] For instance, William Cullen's table had 31 columns, and Torbern Bergman offered up to 59 columns in his many publications.[34] Yet, despite this explosion in substances, Black continued to offer his students a segmented table of around 20 'columns' (which were actually rows, because he had turned them on their side) throughout his career.[35] He did this because he recognised the pedagogical value of using a limited amount of information to unpack complex chemical reactions.

Conclusion

This essay has shown that Black's visualisations were useful learning tools that emerged out of his teaching and research interests. In this sense they were objects that, at the most fundamental level, were designed to be used over and over again and, consequently, they gained meaning through usage. Indeed, they were simple images which he modified so that they could be used in the classroom. Since their simplicity did not visually distract his students, his figures and diagrams functioned as pictures to which he could easily attach various affinity concepts, particularly those that involved the passage of time. Yet, as we've seen, this simplicity was not accidental. It was, in fact, designed.

Black's graphic innovations made his lectures more visually accessible to his students – that is to say, to users who knew relatively little about chemistry. This motivation was probably linked in part to the fact that the number of students who took his course determined his salary. In this respect it literally paid to design teaching aids that reduced complicated topics to simple pictures. Like any informatic tool, the more one used Black's visualisations, the quicker and more helpful they became. This aspect of 'practice makes perfect' reveals an important relationship between pedagogy and what might be called 'informatic time'. The time it took to access the information in Black's visualisations depended on how much time a student had spent copying it into his notes, using it in the classroom and rereading his notes in the privacy of his study. In short, the less experience that a student had with these skills, the longer it took to use and understand the visualisations.

Black used his subtle graphic innovations to reduce complicated chemical theories down to a form of representation that could be understood by students who knew relatively little about chemistry. Key to this understanding was the fact that they could inscribe the visualisation easily and quickly into their notebooks. Likewise, Black did his best to mitigate the visual complexity of a complete affinity table by breaking it up into simple sections that students could easily remember and replicate. These graphic innovations reveal that the acquisition of scientific knowledge in Black's classroom was intimately tied to how it was visualised.

Notes and References

1 For Black's career and the practices of chemical teaching in Edinburgh during the Enlightenment, see Donovan, Arthur L., *Philosophical Chemistry in the Scottish Enlightenment: The Doctrines and Discoveries of William Cullen and Joseph Black* (Edinburgh: Edinburgh University Press, 1975); Eddy, Matthew Daniel, *The Language of Mineralogy: John Walker, Chemistry and the Edinburgh Medical School, 1750–1800* (Aldershot: Ashgate, 2008).

2 Eddy, Matthew Daniel, 'How to See a Diagram: A Visual Anthropology of Chemical Affinity', *Osiris* 26 (2014), pp. 178–96.

3 Eddy, Matthew Daniel, 'Elements, Principles and the Narrative of Affinity', *Foundations of Chemistry* 6 (2004), pp. 161–75.

4 Several aspects of this system are treated in Eddy, Matthew Daniel, 'The Shape of Knowledge: Children and the Visual Culture of Literacy and Numeracy', *Science in Context* 26 (2013), pp. 215–45.

5 An insightful treatment of the visual epistemology that influenced the use of early modern pedagogical diagrams is given in Kusukawa, Sachiko, *Picturing the Book of Nature: Image, Text, and Argument in Sixteenth-Century Human Anatomy and Medical Botany* (Chicago, IL: University of Chicago Press, 2012). See especially ch. 9.

6 Tables can be found in most notes taken in Cullen's lectures. For clear renditions of the mining and furnace diagrams that he used, see Cullen,

William, *Chemical Lectures* (1760), Anonymous note-taker, Bound MS, Wellcome Library, London, MS 1918, ff.126–7, 144–5, 146–7. Examples of the 'lever' diagrams he used to depict chemical reactions are featured in Cullen, William, *Adversaria Chymiæ ex prolectionibus Dr. Gulielm Cullen* (1762), Anonymous note-taker, Bound MS, Wellcome Library, London, MS.MSL.49, ff.44–5. For the pedagogical uses of affinity tables in Britain and France see, respectively, Taylor, Georgette N.L., *Variations on a Theme: Patterns of Congruence and Divergence among Eighteenth-Century Chemical Affinity Theories* (PhD thesis, University College London, 2006), and Lehman, Christine, 'Innovation in Chemistry Courses in France in the Mid-Eighteenth Century: Experiments and Affinities', *Ambix* 57 (2010), pp. 3–26.

7 The reception of Bergman's discovery of ponderous earth in Scotland is addressed in Eddy, *Language of Mineralogy*, pp. 137–44.

8 Black, Joseph, *A Course of Lectures on the Theory and Practice of Chemistry*, vol. 2 (1782), Anonymous transcriber, Bound MS, Royal Society of London, MS/147/2.

9 The pedagogical origins of Black's chiasm and his modification of his chiastic and circlet diagrams are explained in Eddy, 'How to See a Diagram'.

10 Aside from the hundreds of experiments recounted every year in Black's course, other good examples of the step-by-step material manipulations that Black gave his students orally can be seen in the printed 'handouts' that he made as early as the 1760s. See Black, Joseph, *The Preparations of Antimony* (n.d.) and *The Preparation of Mercury Antimony* (n.d.), in Black MS (1766–7), tucked inside vol. 7. Both antimony and mercury were important ingredients for drugs and his students would have valued having copies of how to prepare them.

11 Black, *A Course of Lectures*, lecture 61.

12 Indeed, this pictorial conundrum not only affected how Black's students and contemporaries read notes taken in his lectures, it also affects modern historians who must grapple with the fact that they often cannot see time or space in the diagrams unless they consult the lecture notes written by Black or his students.

13 Experimental or observational procedures played a central role in the graphic management of information in early modern settings. As convincingly argued by Omar W. Nasim, it is difficult to grasp the centrality of these routines without paying very close attention to scribal iterations that take place over time in scientific notebooks. Nasim, Omar W., *Observing by Hand: Sketching the Nebulae in the Nineteenth Century* (Chicago, IL: University of Chicago Press, 2013). See especially his comments in the introduction on procedures.

14 The largest number of watercolour depictions of apparatus that I have encountered appear throughout Black, Joseph, *Lectures on Chemistry*, 6 vols (1778), Paul Panton [note-taker], Bound MS, Chemical Heritage

Foundation, Philadelphia, QD14 .B533 1828. It is unclear whether these depictions were made by Panton, or whether he commissioned an artist to make them.

15 Taylor, Georgette, 'Pedagogical Progeniture or Tactical Translation? George Fordyce's Additions and Modifications to William Cullen's Philosophical Chemistry – Part II', *Ambix* 61 (2014), p. 262.

16 The term 'graphic gropings' is used to represent *in situ* laboratory work in Kemp, Martin, *Visualizations: The Nature Book of Art and Science* (London: University of California Press, 2000), pp. 72–3.

17 Black, *A Course of Lectures*, lecture 61.

18 The importance of avisual forms of evidence in early modern chemistry is underscored in Roberts, Lissa, 'The Death of the Sensual Chemist: The New Chemistry and the Transformation of Sensuous Technology', in David Howes (ed.), *Empire of the Senses: The Sensual Cultural Reader* (Oxford: Berg, 2005), pp. 106–27.

19 A number of Black's instruments are still extant. See Anderson, R.G.W., *The Playfair Collection and the Teaching of Chemistry at the University of Edinburgh 1713–1858* (Edinburgh: The Royal Scottish Museum, 1978). The instrumental context for gravimetric analysis in early modern Scotland is given throughout Connor, R.D., A.D.C. Simpson and A.D. Morrison-Low, *Weights and Measures in Scotland: A European Perspective* (Edinburgh: National Museums of Scotland, 2004).

20 Drawings of instrument pictograms occur regularly in student notebooks taken throughout Black's tenure. For representative examples from different decades see: Black MS (1766–7/1966), pp. xviii, 16, 17,21, 72, 74, 108; Black, Joseph, *Notes from Black's Chymistry* (c.1796), Bound MS, Alexander Monro *Tertius* [note-taker], University of Otago Special Collections, f. 25.

21 Black also summarised collections of related experiments conducted over a considerable period of time outside the classroom. His discussion of what might be seen as a theory of lime takes this approach. In a lecture or two, he summarised a number of experiments that would have taken days to perform outside the classroom. Black, Joseph, *Notes from Dr. Black's Lectures on Chemistry 1767/8*, Thomas Cochrane [note-taker], ed. Douglas McKie (Cheshire: Imperial Chemical Industries, 1966), pp. 68–71.

22 Omitting the visual or verbal representation of experiments or calculations that had contributed to the establishment of a scientific proposition, corollary or principle was a crucial early modern information management technique. Isaac Newton, whose methods served as a guide for most natural philosophers, omitted calculations and experiments 'for brevity's sake' in the later editions of his *Principia*. The mathematical context of these omissions is addressed in Smeenk, Chris and Eric Schliesser, 'Newton's *Principia*', in Jed Z. Buchwald and Robert Fox (eds), *The Oxford Handbook of the History of Physics* (Oxford: Oxford University Press, 2013), pp. 109–65; see especially pp. 151–2.

23 Black, Joseph, *Notes of Dr Black's Lectures, 9 Volumes* (1766–7), Charles Blagden [note-taker], Wellcome Library, London, Bound MS 1219–1227.

24 Blagden's unfinished circles occur in Black MS (1766–7), Notebook 9, f. 634 recto. His square revisualisations occur on f. 634 verso.

25 Watson, Richard, *A Plan of a Course of Chemical Lectures, by R. Watson D.D., F.R.S. and Regius Professor of Divinity in the University of Cambridge* (Cambridge: Archdeacon, 1771). Venel, Gabriel-François, *Cours de Chimie*, ed. Christine Lehman (Dijon: Editions Universitaires de Dijon, 2010). Many examples of eighteenth-century affinity tables occur throughout Duncan, Alistair, *Laws and Order in Eighteenth-Century Chemistry* (Oxford, 1996).

26 Dichotomous tables are also called 'branching tables' in the historical literature. They occur throughout Cullen, Bound MS (1760); see specially ff.102–3 and 152–3. For Black's dichotomous tables, see the three loose-leaf sheets (recto and verso) of dichotomies in Black, Bound MS (1778), vol. 2. As evinced in the work of Georg Wolffgang Wedel's influential *Theoremata Medica* (Jenae: Johannis Bielckii, 1677), dichotomous tables were common in medical chemistry since at least the seventeenth century. For other examples contemporary to Black's teaching, see Dossie, Robert, *Institutes of Experimental Chemistry*, vol. 1 (London, 1759), p. 275; Watson, *A Plan of a Course of Chemical Lectures*.

27 'A TABLE of the Relations or Affinities Observed between Different SUBSTANCES', in Lewis, William, *The New Dispensatory* (London: Nourse, 1753), p. 11. There seems to have been another printing of this book, because the table in the 1753 Nourse edition of *The New Dispensatory* housed in the Bodleian Library, Oxford University occurs on page xi. Copies of both editions are housed on the Eighteenth-Century Collections Online Database.

28 Black MS (1766–7/1966), p. 119. Black directed his students' attention to Lewis's mercury preparations.

29 The logic of Geoffroy's arrangement is discussed in Klein, Ursula and Wolfgang Lefèvre, *Materials in Eighteenth-century Science: A Historical Ontology* (Cambridge, MA: MIT Press, 2007), pp. 147–50.

30 It should be noted here that Panton's affinity table lacks a heading for the fourth grouping on the table. Such a heading, however, is present in most other sets of student notes.

31 Thomas Cochrane's microtables are reproduced in Black MS (1766–7/1966), pp. 161–5.

32 Charles Blagden's version of the affinity table is recorded in Black MS (1766–7), vol. 9., f. 631r.

33 Duncan, *Laws and Order*, pp. 112–14, 132–6.

34 Cullen, William, *Lectures on Chemistry delivered at Edinburgh University, 3 Volumes* (1765), William Falconer [note-taker], Bound MS, Wellcome Library, London, MSS 1919–1921. Cullen's 31-column table occurs in vol. 2, MS 1921. Bergman's tables are discussed in Duncan, *Laws and Order*. See

also the 50-column table, 'Single elective attractions: in the Moist Way; in the Dry Way', in Bergman, Torbern Olaf, *A Dissertation on Elective Attractions* (London: J. Murray, 1785).

35 In his paper on the discovery of carbon dioxide (which he called 'fixed air'), Black argued for the expansion of Geoffroy's affinity table. Black, Joseph, 'Experiments upon Magnesia Alba, Quicklime, and Some Other Alcaline Substances', *Essays and Observations, Physical and Literary*, 2 (1756), pp. 157–225. See especially pp. 224–5. Other late eighteenth-century chemistry teachers also used a simple affinity table for university chemistry lectures. See, for example, the table in Watson, *A Plan of a Course of Chemical Lectures*.

SEVEN

Materia Chemica:
Excavation of the Early Chemistry Stores at Old College, University of Edinburgh

TOM ADDYMAN

Introduction

This paper outlines a discovery of chemistry-related materials made during archaeological excavations within the University of Edinburgh's Old College quadrangle in 2010 and 2011. It concentrates upon the circumstances of their deposition, the range of finds represented and, in advance of the full analytical study that is still ongoing, provides some discussion of their potential significance and their likely association with Joseph Black (1728–1799) and his colleague and successor as professor of chemistry, Thomas Charles Hope (1766–1844).

The Project

Addyman Archaeology undertook an extensive excavation for the University of Edinburgh within Old College quadrangle in 2010, in advance of the laying of a scheme of new paving throughout its interior to a design by Simpson & Brown Architects of Edinburgh.[1] The archaeological investigation involved the excavation of the entirety of the lower level of the quadrangle's area, work that also included a series of localised deeper excavations targeted at the further examination of individual features. The excavation revealed extensive remains of structures and occupation deposits, spanning a period of some 600 years.

Historical Summary

Old College was the site of the medieval church of St Mary (the Kirk o' Field), its associated cemetery and, from the early sixteenth century, a college of priests. Sacked by the English invader in the 1540s, parts of the collegiate precinct were subsequently given over for aristocratic lodging. The site gained notoriety as the scene of the assassination in 1567 of Henry Stuart, Lord Darnley.

In the early 1580s the precinct was chosen by the burgh for the foundation of the 'Tounis College', later to be known as the University of Edinburgh. The College developed piecemeal over the following 200 years to become an extensive 'straggling and irregular' agglomeration of structures, courts and garden areas.[2] Between 1790 and 1830 these were demolished in stages to make way for the existing neoclassical Old College quadrangle buildings.

The arrangement of the early College campus is well documented, earlier sources having been assessed in some detail by Alexander Grant (1884) and more recently by Andrew Fraser (1989). An accurate survey of the complex survives by the surveyor John Laurie in 1767, discussed below, and the appearance of its buildings is depicted in a number of early views such as James Gordon of Rothiemay's of 1647, and subsequently by James Skene of Rubislaw, the Rev. John Sime and others.

The General Excavation

Surviving medieval features on the site, including a part of St Mary's cemetery, were found to be overlain by a number of mid–late sixteenth-century and early seventeenth-century structures, the buildings of the early College. Revealed by the excavation was much of the area of the High College, a spacious court containing many of the institution's principal structures, and parts of the Laigh College, a smaller quadrangle of buildings set at lower level within the north-west quadrant of the site. Along the north side of the High College court were exposed the footings of the south gable wall and kitchen wing of Hamilton House, the mid-sixteenth-century townhouse of James Hamilton, Duke of Châtellerault, a structure retained for the new College. Abutting this on its east side was revealed the remains of the Old Library building of 1642–6 that had occupied most of the remainder of the north side of the court. The footings of the Common Hall building, a very substantial north–south aligned range erected in 1616–17, were exposed along the eastern side of the excavation area; a sunken cobbled courtyard lay along its principal west-facing façade.

The excavations also involved further investigation of individual features and structures by means of a series of more deeply excavated sondages. Four of these were targeted at the rubble-filled interior of the Library building. The footings of its south wall, the principal façade of the structure that fronted on to the High College court, were exposed at points along its length and, at the range's south-west corner, its junction with Hamilton House. It was in the latter area that the find of chemical materials was first made.

The Library Range and its Later History

According to John Laurie's plan of the university complex as it existed in 1767, the Library range was substantial – second only in scale to the 1616–17 Common Hall – and at 25m, of notable length (Colour plate 1). Rothiemay's view of 1647 gave an impression of the range as originally built, notable for its crenellated parapet. In c.1753 the Upper Hall of the 1616–17 range was refurbished and the Library relocated there.[3] In the later eighteenth century the Library range was raised by a further storey, a development well illustrated in a number of early nineteenth-century views. Further plans made during the gradual redevelopment of the quadrangle between c.1790 and c.1830, as well as cartographic sources, show additional details of the range and adjacent structures at that period (Colour plates 2, 3). Finally a view of 1823 by W.H. Lizars, made after the demolition of the range, shows its section in the form of a silhouette against a pre-existing gable wall at its east end.

Combining this early visual evidence, it is clear that by the later eighteenth century the range was a ten-bay structure of three storeys. The range was partly terraced into the natural slope of the ground that descended northwards towards the Cowgate. The lower storey was accessed directly from Printing House Yard on its north side, the third of the early College's courts. Early nineteenth-century views only indicate the presence of two small cellar windows on the south side. At its east end, the lower level incorporated the principal access between the courts, known as the High Transe, which linked the sunken court in front of the Common Hall to the Printing House Yard to the north. Steps at the west end of the frontage gave access to the Library itself, which seems to have occupied the length of the range. The upper storey was added later in the eighteenth century to provide further accommodation.

In 1777 Joseph Black, professor of medicine and chemistry, moved into the Library building, but he soon had to share with John Walker, Regius Professor of Natural History, and the growing Natural History Museum of which the latter was Keeper. In 1780 both professors petitioned the City for the provision of more suitable accommodation. Though Black proposed to use the 1617 Common Hall building, it was decided that a new purpose-built chemistry block be located within Printing House Yard. This was erected in 1781 and is first represented upon John Ainslie's town plan of c.1781. By the time a plan of 1818 was drawn (Colour plate 4), the new block had been connected directly to the lower level of the Library building by an additional link. Both the chemistry block and the Library were demolished in 1820 to make way for the intended north range of William Playfair's revision of Robert Adam's architectural scheme for what is now called the Old College.

Excavation within the Library Range

Three of the four exploratory sondages dug within the Library range extended down to just above the level of the cellar floor. In each case excavation involved removal of a deep deposit of demolition rubble relating to the dismantling of the building in c.1820. Within two sondages sited at the western end of the range, finds relating to its occupation were recovered from the lower parts of this rubble fill and from immediately beneath – where a concentrated level of artefacts was found to directly overlie a dark humic deposit that seemed to represent the upper surface or trample material upon the floor level within the range. The finds included assorted sections of glass tubing (Colour plate 5), some fragments of ceramic vessels (Colour plate 6) and various brightly coloured chemical compounds (Colour plate 7). With the discovery of the latter a clear health and safety issue was raised. Excavation in 2010 was terminated at that point and the compounds sent to the Scottish Environmental Technology Network for identification. Amongst the materials identified were arsenic, and compounds of mercury and cobalt. The highly toxic nature of these substances dictated the termination of archaeological investigation in that area, and the trenches were simply back-filled.

The 2011 Project

Preliminary assessment of the chemical materials and associated finds recovered from the two sondages indicated some of the contents of a chemistry laboratory or stores. The possibility that some of the material may have been associated with the tenure of Joseph Black, professor of medicine and chemistry at Edinburgh from 1766–99[4] and one of the great figures of the Scottish Enlightenment, was raised and was apparently corroborated in part by a reference in Edinburgh's Dean of Guild records of 1800, the year after Black's death, that noted his materials and apparatus were being stored within the lower level of the library range by his successor as professor of chemistry, Thomas Charles Hope.[5]

Further confirmation of the possibility came with preliminary assessment of the ceramics from the sondages, where two types of vessel were identified that seemed certainly to have been products supplied by Josiah Wedgwood to scientist colleagues, including Joseph Black.[6] The material included apparent sections of a high-fired white clay retort, salt-glazed internally, and a vessel with numerical gradations incised into the clay (subsequently identified as sections of the same vessel, an alembic). The small assemblage of material was also assessed by Dr Robert Anderson,[7] who confirmed the probable association of both the glass and ceramic wares with the laboratory activities of Joseph Black. Some of the excavated items appeared to be paralleled in the Playfair Collection in the National Museums of Scotland, Edinburgh[8] – however, the archaeologically recovered material was considered to have a more secure context and historical provenance.

In summary, it was concluded:

- The find was of exceptional importance in terms of the history of science
- The objects individually have very great potential to shed light on eighteenth-century scientific processes (the apparatus recovered; the chemical compounds present; the direct evidence of how they were being used – residues within vessels, crucibles, etc.)
- The find may also shed particular light upon the activities of the pre-eminent scientist Joseph Black
- The find would also provide important new insights into the early development and supply of laboratory equipment, particularly ceramics and glass (some of which appears to be directly associated with Josiah Wedgwood and Archibald Geddes, a glass-maker of Leith, respectively)
- The contextual archaeological information within the excavated area may also provide considerable further insights – the circumstances of deposition, the internal organisation of the store, and distribution of materials and equipment within the store
- There may exist contextual archaeological information within the excavated area in relation to other activities – the earlier use of the

lower level of the 1642 building; pre-existing activity on the site from sub-floor levels; and in relation to activities perhaps suggested by other categories of find

For these reasons, the examination of a more extensive area and the recovery of a larger sample of the materials within the building was considered desirable. Fortunately, by June 2011 the new paving had not yet begun to be laid in the area of the 1640s Library, and the opportunity to re-examine the find-spot was taken; excavation took place between 14 and 30 June.

The Excavation

Following risk assessment and necessary specialist consultation about the handling of the contaminants present on site, the excavation was undertaken under a strict Health and Safety regime.

An initial area of about 2.5m by 3.5m east–west was opened up at the extreme south-west angle of the Library range. Upper levels were removed mechanically, the remainder excavated by hand. The density and obvious importance of the artefacts recovered once the lower rubble had been cleaned off led to the decision to extend the trench further east to obtain a slightly larger sample. The final area excavated measured 2.5m north–south by 5.0m east–west (Colour plate 8).

The importance of the discoveries and potential for further knowledge from the chemical detritus itself dictated that a 100% soil sample was taken of the 'flooring' deposit in the areas excavated – both for analysis and to ensure recovery of smaller artefacts. Sampling was undertaken on the basis of 1m² grid squares so that concentrations of materials could be located by area – there may prove to be some significance in relation to an individual find's relative position.

Overlying Deposits

The rubble infill deposit extended to an overall remaining depth of some 0.6m–0.7m. It consisted of demolition dross formed of smaller pieces of rubble stone within a general matrix of crushed mortar and some wall plaster. An absence of internal stratification suggested the deposit represented the residue of a continuous dismantling operation, a process that must have involved systematic recovery of larger building stone and other recyclable materials such as timber, roofing slates, glass, etc. The resulting debris was backfilled into the lower void of the range's cellar until the ground could be levelled off at the required height as a base for new construction.

A few individual items were recovered from the lower parts of the demolition rubble back-fill – including most of two shouldered storage jars, thick-bodied but with extremely thin bases – perhaps for heating liquids (Colour plates 9, 10); other ceramic vessels; further sections of the white ceramic 'retort'; and most of a large hard paste white-glazed mortar, with residues adhering within. The mortar fragments were spatially separated from one another but proved to be conjoining parts of the same substantial vessel, which had evidently been broken and mixed up within the lower demolition rubble.

Artefact Distribution

Having rapidly dug down to about 0.05m above floor deposits, the remaining overburden was trowelled off by hand to reveal the underlying level – predominantly a very dark grey compacted humic deposit. At the interface between the two an extensive and varied array of finds was revealed, immediately overlying the 'flooring' surface deposits (Colour plate 11). The overlying demolition rubble was carefully removed at this level to reveal the artefact spread *in situ*. All finds were subject to detailed individual mapping and recording prior to removal, information that might prove of importance in any spatial analysis of the artefacts' distribution within the sampled area of the building (Colour plate 12).

It was recognised that the spread of artefacts upon the surface represented a stratigraphic event in its own right. They evidently comprised some of the loose contents remaining within the chamber at the time of its demolition and left strewn about, partly trampled, partly crushed, occasionally intact. The deposit was thus excavated as a discrete level, the finds separated from those of a similar nature recovered from within the surface itself, the latter apparently trampled or otherwise compressed into it.

Finds that were for the most part very obviously chemistry-related were found across the whole excavation area. Particular groups and concentrations were evident in their distribution. At points particular masses of free-blown olive-coloured glass clearly related to individual large vessels (Colour plate 13). In the centre-west part of the trench a mass of smaller items, crucibles, glass rods, etc. were found clustered within a shallow linear depression. In the south-centre-east and south-east parts of the trench were particular concentrations of chemical compounds (Colour plate 14).

The density and diversity of the finds assemblage proved remarkable. Its initial assessment, summarised here, has since been followed by a protracted post-excavation process of residue sampling, cleaning, conservation and expert assessment whose results are still being collated and related to the excavation data.

Glass Objects

A mass of glass finds were made, these often recovered from localised concentrations; in other cases conjoining fragments were widely distributed. A programme of assessment and vessel reconstruction was carried out.[9] The principal items included a variety of laboratory tubes in clear, green and aqua-coloured glass, of various diameters, including one about 4.5cm in diameter and over 50cm in length of deep aqua glass; some retained residues within. There were numerous glass rods of varying section diameter, probably stirring rods, with one complete example (Colour plate 15). Amongst these were a number of apparent thermometer sections, including one with a bulb. Two clear ground glass stoppers were also recovered (Colour plate 16).

Some larger olive green-glass free-blown vessels represented included a large jar (Colour plate 17), at least three broad diameter shallow bowls with cut lips of a type previously unrecorded (Colour plate 18), necks and fragments of at least two mid–late eighteenth-century black glass wine bottles, and assorted bottle bases containing residues. The larger free-blown vessels in particular seem to bear close comparison with items in the National Museums of Scotland's Playfair Collection, the latter thought to date back to Joseph Black's time and possibly manufactured by the Leith glass-maker Archibald Geddes, with whom Black had a close association.

Ceramics

An important and varied assemblage of ceramic vessels[10] was recovered, many clearly related to pyrotechnical and laboratory processes and others more obviously intended for storage of chemical substances. Of the former the tubular fragments initially identified as a possible retort were found to have been part of an alembic, the upper vessel of distilling apparatus (Colour plate 19). This vessel in particular may have been supplied to Professor Black by Josiah Wedgwood.[11] It has incised markings (Colour plate 20), though it lacks the Wedgwood stamp sometimes found on such apparatus.

The mortar fragments were found to constitute an almost complete vessel (Colour plate 21). Various bottle-like jars with broad mouths were recovered (Colour plate 22), some salt-glazed, of a form and fabric without obvious parallel, just possibly of local manufacture and perhaps experimental. Some of these were thick walled but whose remarkably thin bases suggested they were intended for heating liquids. Globular brown-glazed bottles, paralleled by known examples from London and European sources, are probable storage jars for mercury.

There was a notable group of about fifteen small crucibles of a coarse sand-tempered kaolinitic fabric, of a thrown conical thin-walled profile but with upper parts formed to a triangular section (Colour plate 23). These were imports, probably from Hesse, of a well-known type that employed *mullite* for its refractory properties. The group from Old College is notable for the absence of larger examples; perhaps these reflect use for teaching purposes rather than experimental operations on a more industrial scale. In addition to these were various crucible lids and a larger crucible fragment retaining fused smelt residues within.

Also represented were miscellaneous vessels of a more domestic nature, such as blue and white transfer-printed pearlware bowls and creamware mugs (Colour plate 24), that on the basis of residues present had evidently been employed in the preparation of chemical materials.

Metal

Metal objects included a pewter(?) spoon, a number of iron canisters or pots, one possibly zinc-lined, others containing residues, an apparent smelt dribble, two threaded valve keys (Colour plate 25), and part of a lead window came amongst others. The came (a grooved lead strip used to join two pieces of glass) may have been a product of the demolition process. There were also various off-cuts of metal sheeting.

Chemical

There were a number of concentrations of chemical compounds obviously visible at the surface upon excavation, either as spreads of powdered material or solid lumps. Most were of varying bright colours, some were white. There were also the solidified contents of the base of a vessel (crucible or storage vessel, now gone) – this material of creamy white colour. A considerable number of the glass and ceramic vessels retained residues. Elsewhere individual lumps of raw mineral were recovered, for example fragments of galena (lead sulphide). All of these compounds were recovered for further analysis either as samples, or still adhering to finds.

High quantities of mercury, arsenic and cobalt were identified in the samples initially recovered. However the more systematic programme of samples analysis demonstrated an extensive and varied chemical content.[12] Elements identified included zinc, boron, barium, nickel, copper, antimony, bismuth, iron, gallium and lead, amongst others. High levels of thallium were also found in many samples – historically used as rat poison (though it was not chemically characterised until the mid nineteenth century) which may explain its presence.

Other Miscellaneous

Other finds included a number of exotic marine shells and fossilised shells; a fragment of leather; and a whetstone or touchstone (Colour plate 26).

The Cellar Interior

Some evidence was recorded as to the character of the interior of the lower level of the Library. Wall plaster survived upon its interior wall surfaces, this with some suggestion of a white lime-washed finish. At one point the wall plaster along the south wall preserved the impression of a vertical board, edge-set; it is likely that this formed part of a fixed shelving or press arrangement. The surface upon which the finds lay did not have the character of a formally laid floor, rather it was a slightly undulating, compacted level make-up.

While no *in situ* remains of a timber floor were revealed the interior had probably had such a floor structure. This was suggested by the slight gully running east–west about 1.0m from the south wall within which many finds had accumulated. It is possible that this had been a beam slot for a floor joist laid directly upon or partly within the underlying substrate (evidence for similar floor structures were found within the Common Hall building and within other structures in the wider excavation area). The feature may have been considerably trampled during demolition. The other indicator of a timber floor was the abrupt horizontal lower termination of the surviving areas of still-adhering wall plaster. The excavation evidence suggests the likelihood that had there been a timber floor structure which was removed during dismantling of the building.

It is uncertain from the surviving records or archaeological evidence whether the lower chamber(s) of the Library range had been stone vaulted or had had a joisted timber ceiling.

Underlying Deposits and Features

Apparently immediately pre-dating the deposition of the chemistry materials, and somewhat intermixed with them within the trample of the upper sub-floor deposits, were nearly fifty pieces of founders type, the individual characters and spacers employed in page setting in the printing process (Colour plate 27). Between 1754 and 1769 the College Press was situated within this chamber, referred to in the Town Council minutes in 1754 as 'the printing house immediately under the [Low Library]', and run by the Messrs Gavin Hamilton and John Balfour, Printers to the University.[13] An earth-fast setting of four post-pipes located within the excavation area may relate to the former position of a press or an associated apparatus (Colour plate 28). The evidence for printing activity is a further highly significant find, but one that lies outwith the scope of this paper.

Earlier deposits and features underlying the floor make-up related to the construction of the Library range itself in the mid seventeenth century. There was also some indication of previous activity on the site in the form of an apparent footing trench for the wall of a pre-existing structure.

Conclusion of the Excavation

The density of artefacts and the time-consuming nature of the recording process precluded any further extension of the investigation within the eastern parts of the lower level of the Library building. Overall only about 7.25% (some 12.5m^2 of an estimated 172.5m^2) of the cellar interior was excavated and this at its extreme south-west corner. Very clearly the spread of artefacts must have extended considerably beyond the limits of the excavation area, to the north under the existing upper level parapet walkway and to the east. Doubtless much more material still remains undisturbed.

With no further exposure of floor levels elsewhere within the interior and no other indication of its internal arrangements – whether a unified or sub-divided space – it remains uncertain how far the chemistry-related debris may extend or whether the character of its contents may vary. An intriguing possibility is that if the interior had been used for preparation and experimentation there may yet survive more direct evidence of such activities, possibly in the form of fixed equipment, heating apparatus, etc. The remainder of the floor area of the chamber is likely to survive in most areas, the principal exception being a broad 4m wide swathe cut deeply through it on a north–south alignment, this relating to the installation of a service tunnel later on in the nineteenth century.

Discussion

The chemical materials recovered seem to represent some of the stores and equipment of Professor Joseph Black and his successor, Professor Thomas Charles Hope. The find provides an important window upon the activities of these central figures of the Scottish Enlightenment, their teaching practice, the experimentation with which they were involved and in relation to the procurement of materials and equipment. As described, with the post-excavation process still underway this paper presents the interim findings of the excavation and a preliminary analysis of the recovered materials.

The Chemistry Stores (Later Eighteenth to Early Nineteenth Century)

Upon encountering the deposit of chemistry-related artefacts, it initially proved difficult to explain why they were located within a building with no record of chemistry having been taught therein. However, the discovery of a single reference of 1800, recorded that the chemistry materials and apparatus of Joseph Black were being stored within the lower level of the Library range by his successor as Professor of Chemistry, Thomas Charles Hope.[14] The reference placing his apparatus within a building where there is no further record of chemistry having been taught is thus of considerable interest.

It is also unclear why so many artefacts were abandoned within the building when it was demolished in 1820. This is especially true considering the monetary value assigned to such apparatus and the fact that much of it would be purchased as the private property of the Professor of Chemistry. This point is further illustrated by the existence of the Playfair Collection, a large assemblage of historic chemistry apparatus held by the National Museums of Scotland.[15] Gifted to the museum's precursor in 1858 by Lyon Playfair, the then Professor of Chemistry at

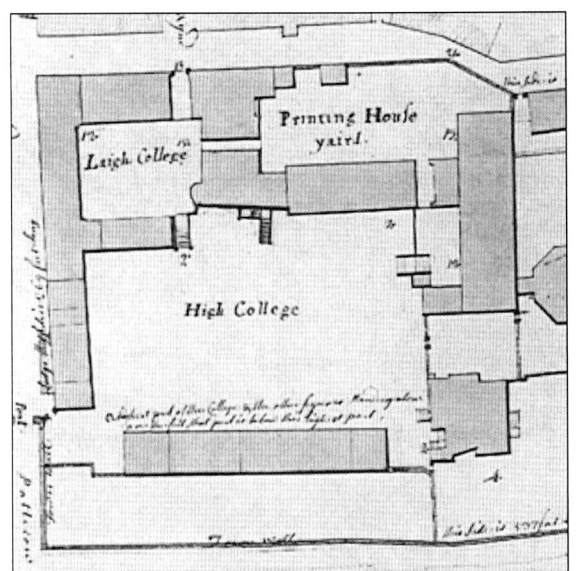

Colour plate 1. *Left*: Plan of the University of Edinburgh in 1767. Drawn by John Laurie (1767). Printing House Yards, site of the chemistry laboratory, is to the top (north).

Colour plate 2. *Below*: 'Museum, Hall and Library of the Old College'. Drawing by James Skene of Rubislaw, 1817.

Colour plate 3. *Bottom*: 'A View of the Inside of the North Part of the College of Edinburgh 1815', lithograph by W. Scott Douglas after an ink and wash drawing by John Sime, reproduced from *Edinburgh in the Olden Time . . . 1717 . . . 1828* (Edinburgh 1880).

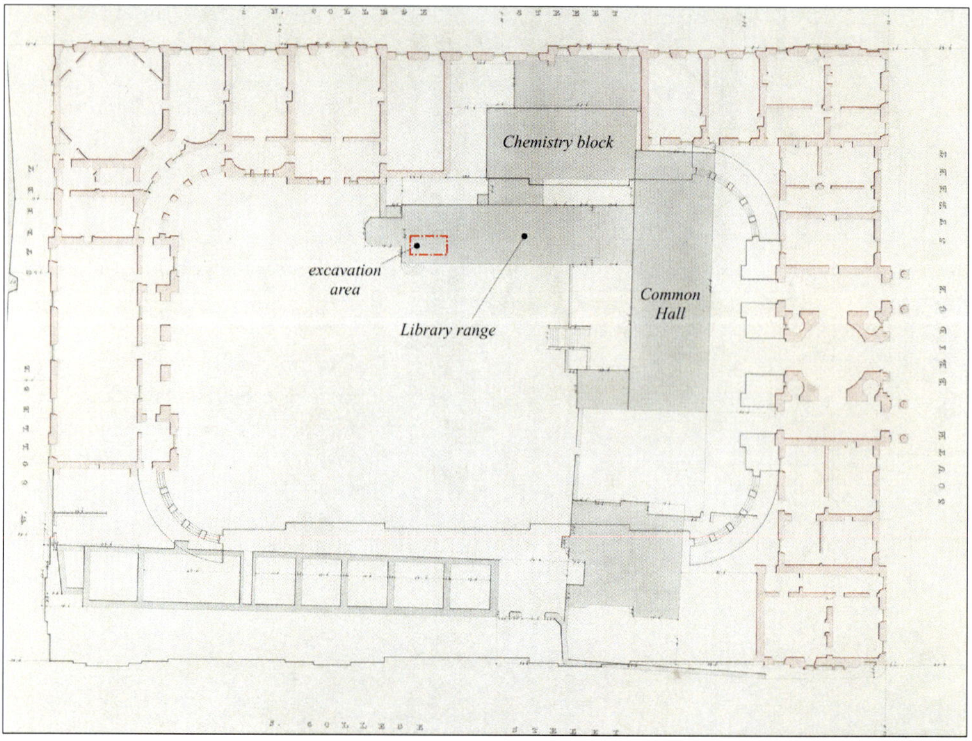

Colour plate 4.
Above: Plan of Old College, William Henry Playfair, c. 1818, with current annotations. The older buildings, not yet demolished, are shown in grey; walls shown in red indicate the Adam/Playfair buildings completed by that date.

Colour plate 5.
Middle right: Glass tubing, one with ground-glass stopper.

Colour plate 6.
Bottom right: Laboratory ceramics, possibly by Josiah Wedgwood.

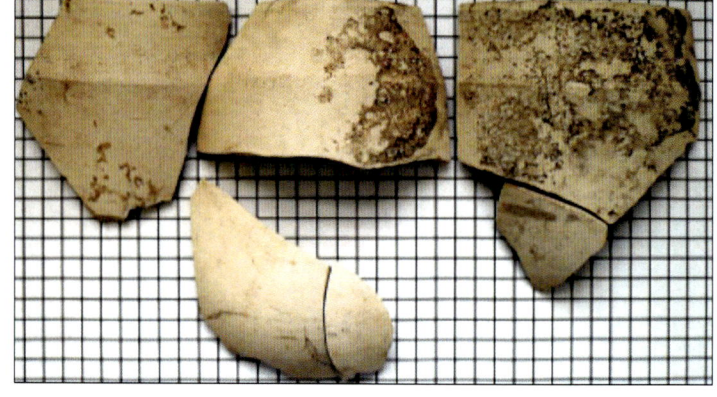

Colour plate 7. *Above*: Chemical compounds uncovered in Old College.

Colour plate 8. *Left*: Extending the earlier trench, looking to the south-east.

Colour plate 9. Ceramic storage jar.

Colour plate 10. Fragments of storage jar, showing very thin base.

Colour plate 11. Detailed recording of object locations, looking south-south-east.

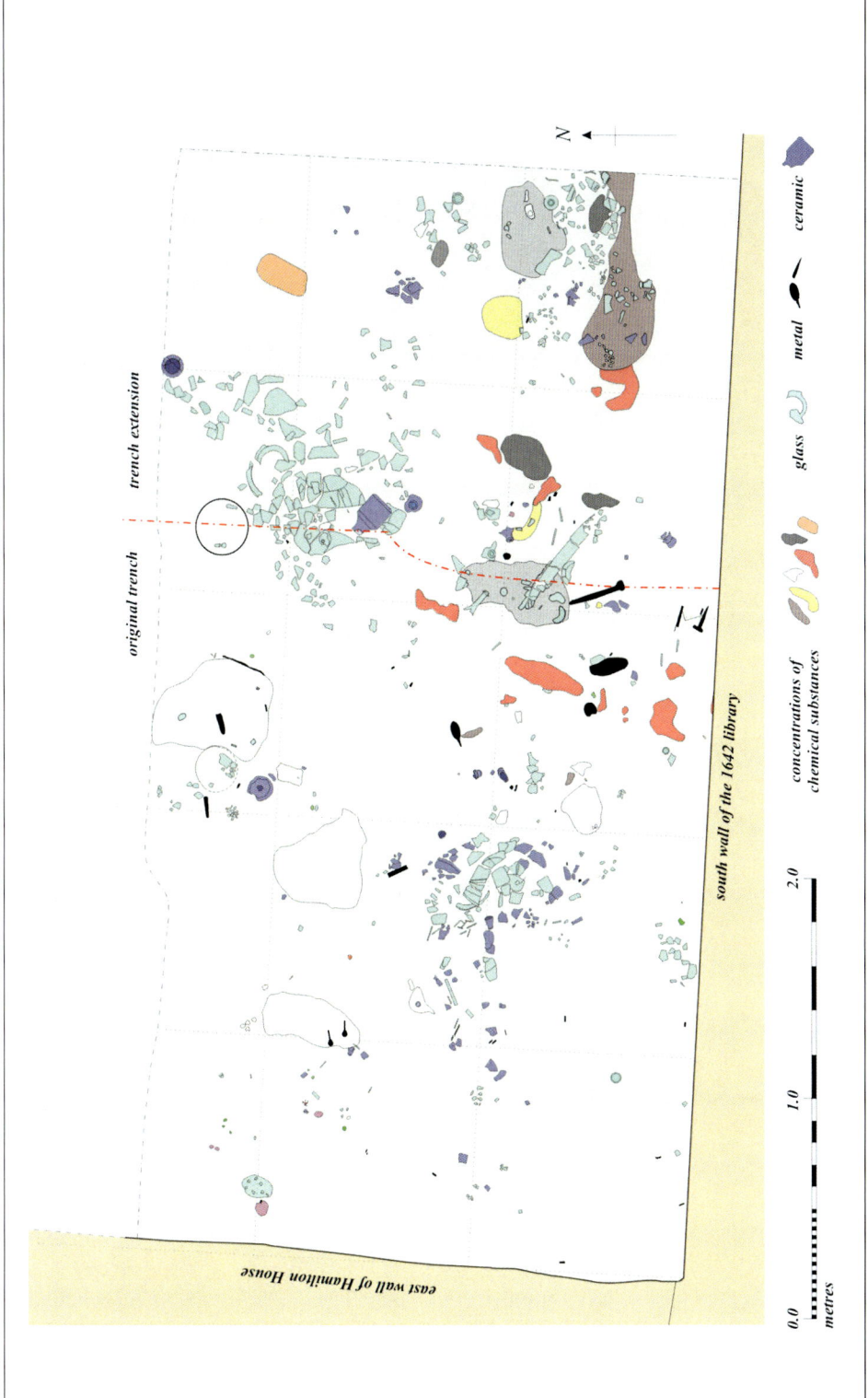

Colour plate 12. Excavation plan showing distribution of object finds and chemical materials.

Colour plate 13. Area of high concentration of finds, including a large fragmented glass object, an iron rim and a large ceramic vessel; to the right, a post-hole.

Colour plate 14. Section of glass tubing, bottle fragments and high concentrations of chemicals.

Colour plate 15. *Above left*: Glass rods.

Colour plate 16. *Above right*: Glass stoppers.

Colour plate 17. *Left*: Large free-blown jar.

Colour plate 18. *Below*: Base of large free-blown glass jar.

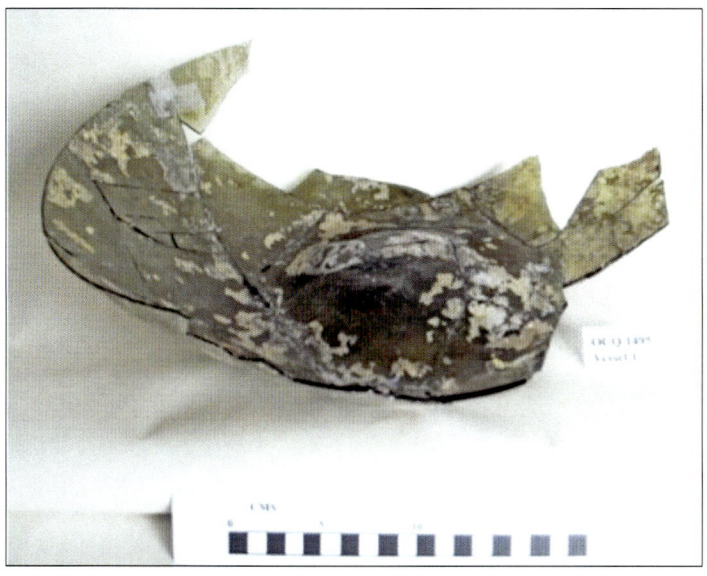

Colour plate 19.
Above: Type of ceramic alembic reconstructed from parts.

Colour plate 20.
Right: Incised markings on the alembic shown in Colour
plate 19.

Colour plate 21.
Below: Large ceramic mortar.

Colour plate 25.
Top: Metal threaded valves.

Colour plate 26.
Above: Touchstone for estimating gold content in alloys.

Colour plate 27.
Right: Piece of founders' type: the letter 'M'.

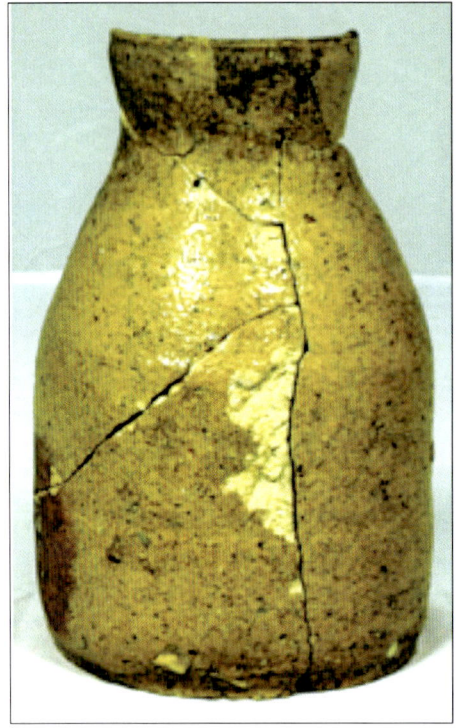

Colour plate 22. Stoneware bottle.

Colour plate 23. Hessian crucible.

Colour plate 24. Blue-and-white cup in Chinese style, probably English.

Colour plate 28. Post-excavation view of trench showing four post-holes, looking south.

Colour plate 29. Joseph Black lecturing in Edinburgh during the 1767–68 session, his assistant behind. Ink sketch by a student, Thomas Cochrane.

Colour plate 30. Joseph Black lecturing in Edinburgh. Etching by John Kay, 1787.

Colour plate 31. William Dean's Class Card for Joseph Black Chemistry Lectures, dated 25 October 1769.

Colour plate 32. Chemical flasks from the Playfair Collection, probably made in Leith glassworks, late eighteenth century.

Edinburgh University, it comprised chemistry equipment held by the Chemistry Department which was then regarded as obsolete or was unused. Many of these objects reputedly date to the tenure of Joseph Black in the second half of the eighteenth century and their preservation within the chemistry department indicates their value and status as working equipment to some extent in the intervening period.

On reviewing the finds assemblage it was felt that, the building having been demolished in 1820, these materials remaining in the cellar may have been those Professor Thomas Hope did not want in his temporary laboratory.[16] He had certainly inherited Black's equipment: he told the Commissioners visiting the Scottish Universities in 1828, 'Dr Black left a very excellent Apparatus, and a considerable collection of mineralogy, which, from a private arrangement, became my property' [i.e. Hope had bought it from Black, presumably between the autumn of 1796 and December 1799]. The material within the stores must legally have been Hope's.[17] It was also felt that while the majority of the assemblage was more likely to be of eighteenth rather than early nineteenth century date, that some items might even pre-date Black's tenure – the form of the fragments of the large mortar perhaps being somewhat earlier.[18]

In terms of comparative material from excavation sites this find has few parallels, and in particular from so apparently secure a context. Of these few perhaps the most significant was a refuse deposit excavated behind the original Ashmolean Museum in Oxford from which were recovered crucibles and covers that closely parallel finds at Old College, and other ceramic vessels of early to mid-eighteenth century date.[19] It will be essential to compare the excavated assemblage to the Playfair Collection held by the National Museums of Scotland (which includes a number of items from Black's laboratories). Certainly the ceramics may be compared with other items provided by Josiah Wedgwood for chemical experimentation, and the glass finds with known products of the Leith Glassworks under the tenure of Archibald Geddes, who was a very likely supplier to Joseph Black.[20]

In addition to the chemistry-related materials were the unexpected finds of what appear to be a number of natural history specimens and an assortment of pieces of founders' type employed in printing. The former most likely relate to the natural history collections housed in the upper levels of the building, and may have fallen from above during the demolition process. The more numerous printing finds seem most likely to relate to activity within the cellar interior itself. Though somewhat intermixed with the chemical materials, the pieces of type (perhaps small enough to fall between floorboards) may have accumulated upon the floor before the area became the chemistry store.

Commentary, 1789

Joseph Black's own commentary of 1789, upon the future requirements for the accommodation of the professor of chemistry, provides a valuable insight into the operation of his existing laboratory:

> [The professor of chemistry] has it is true only one hour of teaching but he must spend several hours every day in his laboratory in preparing for his experiments & operations of the next lecture or in finishing those already begun and as these operations often last ten, twelve, or twenty four hours, or some of them several days, he is under the necessity of looking into it frequently during the day & occasionally must be there early in the morning and late at night. Nor is this sort of labour confined to the Session of the College, he must have recourse to his Laboratory at all times to carry on his Studys and qualify himself the better for the discharge of his duty. To study books he can go to his library but he must have recourse to his laboratory when he studys nature, when he wishes to ascertain or to discover new facts. His Office is much more laborious than [the other medical professors] who have only one hour of lecturing dayly; it is also attended with considerable expence for fewel, furnaces, Glasses & materials.[21]

Contemporary representations

Contemporary illustrations exist of Joseph Black that throw some light upon the find of the chemistry stores and its contents: an ink sketch by a student attending Black's lectures in the 1767–8 (Colour plate 29), session, an oils-on-canvas portrait by David Martin of 1787 (front cover), and an etching of 1787 by John Kay (Colour plate 30). Each shows Black engaged in teaching and in each instance are included items of his attendant apparatus – glass tubes, a flask, a bell jar, etc.

Further visual evidence comes from one of Black's surviving engraved lecture cards, dated October 1769 (Colour plate 31). This illustrates minerals, medicinal plants and pieces of chemistry apparatus such as flasks, distillation vessels and, notably, a crucible of precisely the same form as a number of those recovered from the excavation.

Conclusion

Clearly the 2011 excavation confirmed the importance of the initial find of chemistry materials first identified in excavation trenches in 2010, and a far more comprehensive additional sample of these has now been obtained. The artefacts recovered exceeded expectations in terms of their concentration, diversity and quality of preservation. Their presence seems certainly to confirm that the lower parts of the Library building, at least to the west, were employed for the storage of chemical materials – though this may not have been its only function.

The interim finds assessments demonstrate the extreme interest of the assemblage and the considerable potential for a number of areas of follow-on research. Overall the assemblage has great significance for precisely the reasons already outlined above, and the results of the ongoing programme of post-excavation research will certainly constitute an important contribution to both the archaeology and history of science, and in relation to the experimental activities of one of chemistry's most prominent eighteenth-century proponents.

The chemistry-related finds assemblage and other materials are presently subject to a combined programme of post-excavation sampling, analysis and research. This research is being accompanied by a more detailed survey of documentary sources in relation to the find and its physical and historical context. It is intended that the results of the investigation of the chemistry stores be published in the near future, and will accompany a general account of the wider excavations at Old College (Addyman et al., forthcoming).

Notes and References

1   The project was managed for the University of Edinburgh by George Boag, Department of Estates and Buildings, and overseen by Professor Mary Bownes, Senior Vice Principal; the excavation was supervised by Ross Cameron and Kenneth Macfadyen of Addyman Archaeology.

2   Grant, A., *History of the University of Edinburgh*, vol. 2 (London: Longmans, Green & Co., 1884), p. 184.

3   Ibid., p. 171.

4   Black remained professor until his death in December 1799; however, he had stopped teaching in the mid-1790s, this task being taken up by his

conjoint professor, Thomas Charles Hope. Robert Anderson, pers. comm.

5 Cited by Fraser, A., *The Building of the Old College* (Edinburgh: Edinburgh University Press, 1989), p. 44; City of Edinburgh, Dean of Guild Records: ECA, bundle 16, shelf 36, bay C.

6 Assessment by George Haggarty.

7 Thursday 9 June 2011, at Addyman Archaeology's Leith office.

8 Catalogued by Anderson, R.G.W., *The Playfair Collection and the Teaching of Chemistry at the University of Edinburgh : 1713–1858* (Edinburgh: Royal Scottish Museum, 1978).

9 Assessment by Robin Murdoch.

10 Assessed by George Haggarty, pers. comm.

11 Chaldecott, J.A., 'Wedgwood's Ceramic Wares for Chemical Use: Production and Supply 1779–1794', *Ambix* 28 (1981), pp. 182–205; Science Museum, London, *Josiah Wedgwood: 'the Arts and Sciences United'* (Burslem: Josiah Wedgwood & Sons Ltd., 1978).

12 Initial assessment of some samples by the Scottish Environmental Technology Network; general analysis by Zucana Gojdosechava, under the guidance of Dr Andrew Alexander, University of Edinburgh.

13 Wynkin de Worde Society, *Edinburgh and its College Printers* (Edinburgh: Edinburgh University Press, 1973), pp. 16–17.

14 Cited by Fraser, *The Building of the Old College*, p. 44; City of Edinburgh, Dean of Guild records: ECA, bundle 16, shelf 36, bay C.

15 See Morrison-Low, A.M., 'Surviving Eighteenth-century Chemical Apparatus', this volume.

16 This is speculation, but likely correct. There can be no certainty that the items discovered were not already in a fragmented state when the demolition occurred, and that they would have been of no use to Hope.

17 Robert Anderson, pers. comm., 23 August 2011.

18 Ibid.

19 Hull, G., S. Hamilton-Dyer, P. Blinkhorn and P. Cannon, 'The Excavation and Analysis of an Eighteenth-Century Deposit of Anatomical Remains and Chemical Apparatus from the Rear of the Ashmolean Museum (now the Museum of the History of Science)', *Post-Medieval Archaeology* 37 (2003), pp. 1–28.

20 Anderson, *The Playfair Collection*. Interestingly, there are no ceramic items in the Playfair Collection (and never have been). In all likelihood, Lyon Playfair felt that such items were still of value and did not pass them over to the Museum.

21 Fraser, *The Building of the Old College*, p. 111; Anderson, Robert G.W. and Jean Jones, *The Correspondence of Joseph Black* (Farnham: Ashgate, 2012), vol. 2, pp. 1068–70.

EIGHT

Surviving Eighteenth-Century Chemical Apparatus in the National Museums of Scotland

A.D. MORRISON-LOW

On 11 June 1858, the *Caledonian Mercury* devoted a column of newsprint to an abstract from a document entitled 'Directory of the Industrial Museum of Scotland and of the Natural History Museum, Edinburgh', issued by the Science and Art Department of the Committee of Council on Education. It noted that the Industrial Museum had been established in 1854; that the intended site was in close proximity to the University of Edinburgh; and that in 1857 the Department of Art and Science was transferred from the Board of Trade to the Privy Council Committee on Education, under which the Industrial Museum was placed.[1] But enough of the legal and administrative niceties of establishing a new national museum: some of us have lived through this sort of upheaval, and the more object-minded amongst us are mainly concerned with the collections.

The Industrial Museum of Scotland was one of the ancestor bodies of today's National Museums of Scotland: and as such, whence its historic industrial collections derive.[2] And further on in the newspaper article is a sentence which today's curators in Chambers Street would do well to regard: 'The Industrial Museum of Scotland is not intended to be a Museum of Scottish industry alone, but a Museum of the industry of the world in special relation to Scotland.' Dr George Wilson, its first director, was also appointed the first (and only) Regius Professor of Technology in the University of Edinburgh.[3] He is a fascinating figure; but Robert Anderson has produced many more insights about him than can be described in this short piece.[4] Anderson's catalogue of the Playfair Collection recounts in much more detail just how the University's historic chemical apparatus was acquired, used, fell into disuse and was presented to the Industrial Museum of Scotland by Lyon Playfair in 1858, after Wilson's appointment to the Edinburgh chair and before his early death in 1859.[5]

Playfair and Wilson were both chemists, and Anderson suggests that they may have met in 1837 at Edinburgh, where they were both members of the same student brotherhood. The following year they were both assistants to Thomas Graham, professor of chemistry at University College, London.[6] Part of the importance of the Playfair Collection, as Anderson points out, is that it is one of the very few collections of surviving early chemical material that has retained its provenance (but this dates only from its entrance into the museum's collections, on 30 September 1858); and unlike other eighteenth-century chemistry collections – he mentions Lavoisier's at Paris and van Marum's at Haarlem – the quality of craftsmanship 'is of a significantly humbler standard'.[7]

There are some 75 items in the Playfair Collection, and many of these are associated with the later professors of chemistry: items connected with Joseph Black's immediate successor Thomas Charles Hope, who held the chair until 1843; and with his successor, William Gregory, the incumbent between 1843 and 1858, son of a famous scientific Scottish family, but whose own interests lay in the rather more dubious enthusiasms of phrenology and mesmerism.[8] One intriguing piece made from a triangular plate of copper inset with a glass sphere, probably owned by Gregory and identified by Wilson as a 'tetragrammaton' on entry to the museum's collections, is the only item in the Playfair Collection which (as Anderson says) 'has no connection with conventional laboratory operations'.[9] This piece, along with other early nineteenth-century parts of the Playfair Collection, will not be discussed in this paper.

There are, of course, a number of iconic pieces in the collection containing little or no glass, including an early seventeenth-century Hauksbee-type pump, being one of only eight examples known to survive.[10] Possibly inspired by its presence, George Wilson wrote a paper 'On the early history of the air-pump in England', although the provenance of the pump at Edinburgh remains unclear.[11] Another of the museum's most treasured objects is Joseph Black's balance, about which there has been much discussion.[12]

The main purpose of this paper, however, is to discuss the laboratory glassware in the Playfair Collection (Colour plate 32). Altogether there are 22 surviving pieces, and their history and provenance may shed some light on the endeavours of Tom Addyman and his team, who spent much of the summer of 2011 digging in the foundations of the site of Joseph Black's laboratory, finding shards of pottery and glassware that were buried there. Did these glass and pottery slivers come from the same sources as the surviving pieces? Did they have a local origin, or come from farther afield?

Archibald Geddes was an Edinburgh glassmaker, who died in 1809.[13] He had attended Joseph Black's lectures for 1778–9 and 1779–80, and became friendly with the professor of chemistry.[14] Black, of course, was interested in glassmaking, both as an industrial and chemical process, but also as an investment.[15] At a time when the glass trade was expanding in Edinburgh and Leith, Leith specialised in the manufacture of wine bottles, largely for export to France and Spain (Plate 7). At its peak in around 1770, production was a staggering one million bottles per week. The Leith-pattern bottle is the parallel-sided, round shouldered, narrow-neck bottle now dominant within the wine industry.[16] Made from brown or green glass, it is not a great stretch of the imagination to see how the basic material (and shape) could be made into something for the lecture demonstration.

Geddes was first manager, then partner and subsequently (with his son) owner of the Edinburgh Glasshouse Company at Leith. In 1794 Black made a loan of £500 to Archibald and William Geddes to make improvements, and Black became a director of the company. At the pinnacle of its success the company made flint, crown and bottle glass, and some of this was used to make scientific apparatus.[17] An account of 1792 states:

> About thirty years ago there was only one glass house company in Scotland [. . .] there are now six glass houses at Leith alone. At the time I first mention, nothing less than bottles of coarse green glass was made there; and to that article the glass house company at Leith confined their efforts, till about a dozen years ago, when they began to make fine glass for phials, and other articles of that nature [. . .][18]

Among the 'other articles' was a glass device for carbonating water. Joseph Priestley had suggested this in the early 1770s, but John Nooth, who had studied medicine under Black at Edinburgh, graduating MD in 1766, read a paper describing an improved version in December 1774 (Plate 8).[19] 'Fixed air' was Joseph Black's term for carbon dioxide. Nooth's apparatus for absorbing fixed air in water consisted of three glass vessels fitted together with airtight joints. Fixed air generated in the lowest vessel passed through a valve into water contained in the middle vessel. Any water displaced upwards by the gas entered the top vessel. The apparatus, which made possible the domestic production of 'spa water', was commercially successful, and large numbers were manufactured (although because of its fragility, not many complete examples have survived).[20]

Archibald Geddes appears to have been behind the production of a pamphlet advertising his version of this device – 'aerated medicinal waters, by means of the improved glass machines' in 1787, with recipes for 'alkaline aerated water', 'selters water', 'pyrmont water' (Priestley's phrase) and 'spaw water'.[21] In the back of the pamphlet is an advertisement, showing that Geddes's glasshouse was clearly capable of relatively sophisticated productions. From 1787, the *Edinburgh New Dispensary and Pharmacopeia* (also 'printed for William Creech') advertised pharmaceutical glass measures made by the Edinburgh Glass-house Company of Leith. However, perhaps the main reason for emphasising the importance of Archibald Geddes is that from being a formal pupil of Joseph Black, he became a close friend: to the extent that John Robison wrote in the Preface to his 1803 edition of Black's *Lectures on the Elements of Chemistry*, that:

> Soon after his coming to Edinburgh [. . .] Doctor [Black] got another pupil, Mr Archibald Geddes, manager of the glass-works at Leith, who soon engaged his Professor's attention by the readiness and propriety with which he applied to the improvement of his manufacture the instructions which he received in the lecture. Farther acquaintance shewed more to esteem and attach; and it terminated in the most intimate and confidential friendship. From this friend no circumstances of Dr Black's former life or present condition was withheld; and to his assistance he had recourse in every thing that affected either his fortune or his comfort.[22]

Such was their friendship, that Geddes was named one of the executors of Black's estate.[23]

Of course, Geddes was not the sole manufacturer of glassware, especially where instruments were concerned. Alexander Wilson, who became professor of practical astronomy at Glasgow in 1760, made thermometers for himself and colleagues, including Joseph Black: an example was acquired by the University of Edinburgh at an unknown date, although it has a manuscript date on the scale of '1782'.[24] Wilson, like Joseph Black, has been identified by Roger Emerson as one of the placemen of Archibald Campbell, Earl of Ilay, but they seem to have known each other independently.[25] Wilson was originally from St Andrews, and after a London apprenticeship as a surgeon and apothecary, became a type founder with connections to the renowned Foulis Press. He had been making instruments since the 1730s, when he assisted his fellow-townsman George Martine in his experiments on heat; in 1740, Martine wrote that thermometers were 'made nowhere in greater perfection, or with greater exactness, than by our countryman *Wilson* at *London*'.[26] Wilson is first mentioned in Black's correspondence in connection with thermometer-making in 1753, and by early 1768 he was writing:

Dear Doctor, I have sent in to Edinburgh some small pocket Thermometers if any of your Students want them please inform them that they are to be had of Messrs Auld & Smellie at their printing house in the luckenbooths [...][27]

A final example of a maker of chemical apparatus from the eighteenth century is the famous potter, Josiah Wedgwood.[28] He invented the Wedgwood pyrometer (Plate 9) for measuring the high temperatures in kilns, and made chemical apparatus, some of which he gave to friends, including Black, who taught Wedgwood's sons at Edinburgh.[29] The collections at the National Museums once held over 1,000 examples of a collection of specimens illustrating the manufacture of Wedgwood pottery from Etruria, Staffordshire, much of which came into the collection in 1856 and 1859. Collected by George Wilson to demonstrate production processes, a proportion of this material survives, including an example of the Wedgwood pyrometer.[30] It was not until 1974 that my former colleague, Dr Anderson, acquired a glazed earthenware chemical retort made by the firm and stamped with its name in about 1800.[31] No chemical apparatus from Josiah Wedgwood's Etruria was passed on by the University to become a part of the Playfair Collection, which makes the pieces dug out of the ground two summers ago all the more intriguing.[32]

Notes and References

1 *Caledonian Mercury*, 11 June 1858, quoting *Directory of the Industrial Museum of Scotland* (London: George E. Eyre, 1858), pp. 5–6.

2 For histories of this institution, see Allan, D.A. *The Royal Scottish Museum 1854–1954* (Edinburgh: Oliver & Boyd, 1954); Calder, J., *The Royal Scottish Museum* (Edinburgh: Royal Scottish Museum, 1984); Calder, J. (ed.), *The Wealth of a Nation* (Edinburgh: National Museums of Scotland, 1989) and Yanni, C., *Nature's Museums: Victorian Science and the Architecture of Display* (London: Athone Press, 1999), ch. 4: 'Nature as Natural Resource: The Edinburgh Museum of Science and Art', pp. 91–110.

3 Hartog, P.J., rev. by R.G.W. Anderson, 'Wilson, George', in *Oxford Dictionary of National Biography* (Oxford: Oxford University Press, 2004).

4 Anderson, R.G.W., '"What is Technology?": Education through Museums in the Mid-Nineteenth Century', *British Journal of the History of Science* 25 (1992), pp. 169–84; Anderson, R.G.W., 'Connoisseurship, Pedagogy, or Antiquarianism?', *Journal of the History of Collections* 7 (1995), pp. 211–25.

5 Anderson, R.G.W., *The Playfair Collection and the Teaching of Chemistry in the University of Edinburgh, 1713–1858* (Edinburgh: Royal Scottish Museum, 1978), pp. 57–62.

6 See Hartog, 'Wilson, George'.

7 Anderson, *The Playfair Collection*, p. 1. For Lavoisier's apparatus, see Holmes, Frederic, 'The Evolution of Lavoisier's Chemical Apparatus', in F.L. Holmes and T.H. Levere (eds), *Instruments and Experimentation in the History of Chemistry* (Cambridge, MA: MIT Press, 2000), pp. 137–152; for that of van Marum, see Forbes, R.J., E. Lefebvre and J.G. de Bruijn (eds), *Martinus van Marum: Life and Work* (Dordrecht: Springer, 1980); Turner, G.L'E. and T.H. Levere, *Van Marum's Scientific Instruments in Teyler's Museum*, vol. IV (Leiden: Noordhoff International, 1973).

8 Anderson, *Playfair Collection*, pp. 65–6, where Anderson associates each item with one or another of the professors.

9 Ibid., p. 123.

10 Brundtland, T., 'Francis Hauksbee and his air pump', *Notes and Records of the Royal Society* 66 (2012), pp. 253–72. The pump's original glass receiver does not survive.

11 Wilson, G., 'On the early history of the air-pump in England', *Proceedings of the Royal Society of Edinburgh* 2 (1849), pp. 207–14.

12 Anderson, *Playfair Collection*, pp. 73–5; Calder, *Wealth of a Nation*, pp. 60–1; Newcomb, S., *World in a Crucible: Laboratory Practice and Geological Theory at the Beginning of Geology* (Boulder, CO: Geological Society of America, 2009), p. 83.

13 *The Scots Magazine* 71 (1809), p. 239.

14 Anderson, R.G.W. and J. Jones (eds), *Correspondence of Joseph Black* (Farnham: Ashgate, 2012), vol. 2, p. 1408.

15 Ibid., pp. 1206–7. See also Anderson, Robert G.W., 'Who was the Real Joseph Black?', in G.D. Patterson and S.C. Rassmussen (eds), *Characters in Chemistry: A Celebration of the Humanity of Chemistry* (Washington, DC: American Chemical Society, 2013), pp. 37–47.

16 Grant, J. (ed.), *Cassell's Old and New Edinburgh* (London, Paris and New York: Cassell, Petter, Galpin & Co., 1883), vol. 3, p. 239. A number of bottles for domestic use associated with Leith are in the collections of the National Museums of Scotland, among them: H.MEN 43: Flagon of dark olive glass with brass stopper and ring, and bottle stamp marked 'DM CUSTOMS', from Leith, late eighteenth to early nineteenth century; H.MEN 45 A: Bottle of olive glass with straight sides, from Lamb's Building, Waters Close, Leith, early nineteenth century; H.MEN 73: Bottle of dark brown green glass, with a globular shape and a round seal 'R.H 1826'; reputedly from Leith; H.MEN 230: Wine bottle of dark green glass with extensive chip engraving and marked 'Blown at Leith NB' on the base; A.1894.448: Bottle of dark green glass with scratched design of crown, rose and thistle: Scottish, Leith, c.1820.

17 Anderson, *Playfair Collection*, p. 143 and p. 146; Anderson and Jones, *Correspondence of Joseph Black*, p. 54, p. 542 n. 4, p. 1207 n. 5 and p. 1457. Douglas, R.W. and S. Frank, *History of Glassmaking* (Henley-on-Thames: Foulis, 1972), p. 30, suggest that Geddes was an able industrialist. Fleming, A., *Scottish and Jacobite Glass* (Glasgow: Jackson, Son & Co., 1938), p. 112, gives some background: Geddes was appointed by the Edinburgh and Leith Glass Company, which installed modern glassworks in Salamander Street. Although he came from a landed Galloway family, a number of his close relatives became distinguished glassmakers. The convoluted history of the genesis of the Leith glassworks to the mid eighteenth century is recounted by Turnbull, J., *The Scottish Glass Industry 1610–1750: 'To Serve the Whole Nation with Glass'* (Edinburgh: Society of Antiquaries of Scotland Monograph Series No 18, 2001), pp. 145–74.

18 [A., J.] [Anderson, James], 'Hints respecting the progress of manufactures and their present state in Scotland', *The Bee, or Literary Weekly Intelligencer* 10 (1792), p. 333.

19 Priestley, J., *Directions for Impregnating Water with Fixed Air; in Order to Communicate to it the Peculiar Spirit of Pyrmont Water and Other Mineral Waters of a Similar Kind* (London: J. Johnson, 1772); Nooth, J.M., 'The description of an apparatus for impregnating water with fixed air', *Philosophical Transactions of the Royal Society* 65 (1775), p. 59.

20 Examples survive at the National Museums of Scotland, T.1974.212; the Science Museum, London, inv. no. 1982–534; and at the National Trust property Cragside, Northumbria: see http://ntcragsidehouse.wordpress.com/2013/09/15/object-in-focus-nooths-apparatus/ (last accessed 30 December 2013).

21 Anon., *Directions for preparing aerated medicinal waters, by means of the*

improved glass machines made at Leith Glass-works (Edinburgh: Printed for William Creech, 1787).

22 Black, J., *Lectures on the Elements of Chemistry, Delivered in the University of Edinburgh . . . Published by John Robison* (Edinburgh and London: A. Constable and T.N. Longman and O. Rees, 1803), p. vi.

23 Anderson and Jones, *Correspondence of Joseph Black*, p. 1460, pp. 1467–9.

24 For Wilson, see Stronach, G., 'Wilson, Alexander', rev. by R. Hutchins, in *Oxford Dictionary of National Biography* (Oxford: Oxford University Press, 2004), http://www.oxforddnb.com/view/article/29633?docPos=2 (last accessed 30 December 2013). The thermometer at the National Museums of Scotland is T.1975.56; there is another at the Science Museum, London, inv. no. 1954–260.

25 Emerson, R., 'Politics and the Glasgow Professors', in A. Hook and R.B. Sher (eds), *The Glasgow Enlightenment* (East Linton: Tuckwell Press, in association with Eighteenth-Century Scottish Studies Society, 1995), pp. 21–39 (esp. p. 31). Wilson's instrument-making is discussed by Morrison-Low, A.D., '"Feasting my eyes with the view of fine instruments"', in C.W.J. Withers and P. Wood (eds), *Science and Medicine in the Scottish Enlightenment* (East Linton: Tuckwell Press, 2002), pp. 18–53.

26 Martine, G., *Essays Medical and Philosophical* (London: A. Millar, 1740), p. 27.

27 Quoted in full in Anderson and Jones, *Correspondence of Joseph Black*, pp. 195–8 (on p. 195).

28 Reilly, R., 'Wedgwood, Josiah', in *Oxford Dictionary of National Biography* (Oxford: Oxford University Press, 2004); online edn, September 2013, http://www.oxforddnb.com/view/article/28966 (last accessed 30 December 2013).

29 Anderson and Jones, *Correspondence of Joseph Black*, pp. 1431–2.

30 National Museums of Scotland T.1856.89.601.

31 National Museums of Scotland T.1974.69; purchased from a London dealer.

32 See Addyman, Tom, 'Materia Chemica: Excavation of the Early Chemistry Stores at Old College, University of Edinburgh', this volume.

NINE

Joseph Black's House in Nicolson Street:
Its History and Ultimate Fate

PETER J.T. MORRIS

The Historiographical Problem

This paper solves a long-standing issue of where Joseph Black's final house (and place of death) was located, what it looked like and what its ultimate fate was. It is well known that it was in Nicolson Street, which runs south out of Edinburgh, and that it was taken over by the Asylum for the Industrious Blind (later the Royal Blind Asylum). However, there were two blind asylums in the same street, and they are rarely distinguished when Black's house is mentioned, which has given rise to confusion. I erroneously stated in an article in 2004 that Black's house could be the current 48 Nicolson Street, but this is actually the building next door, which dates from the same period.[1] As well as clearing up a minor historical mystery, this account of Black's house also serves as a reminder that the homes of famous scientists are easily forgotten and rarely preserved.

Joseph Black's Homes in Edinburgh

When Black moved from Glasgow to Edinburgh in 1766, he lived in a house on the east side of the ancient College Wynd, overlooking the ornate entrance to the Tounis College (Town's College was the usual name given at the time for the University of Edinburgh), which no longer exists.[2] Eight years later, he moved to a house in Argyle Square, which was eventually demolished in the 1860s to make way for Chambers Street. Finally, in 1781, Black sold this house to his nephew-in-law Adam Ferguson, the historian, but he only lived there for five years.[3] Black then bought a new house built by James McPherson in Nicolson Street for £1,470.[4] He died here in 1799. In the summer he often moved to houses elsewhere, but always in the Edinburgh area, to remain close to his laboratory and friends. He rented Higher Hermitage on Leith Links (now replaced by a group of small streets south of Leith Links)[5] in the summer of 1794, and in the summer of 1798 and again in 1799 he rented a house on the Meadows, in Sylvan Place. This is the only house associated with Black which may still exist, but its identification as such remains uncertain.

The House in Nicolson Street

Black's will gives a good description of the building:

> large dwelling house with the balustrades in the front consisting of three floors and garrets and four vaulted cellars below the pavement opposite to the front of the house. Together with the back court of the said house with the stable coach house [working] house pump well and others thereon, all lately built by James McPherson mason and architect at Dean near Edinburgh upon and consisting of the southernmost part or division of that plot or piece of ground in Nicholson park feued by him from Thomas Carnegy of Craigs Esqr.[6]

This is identical to the description given in the sasine of Black's house after it was sold.[7] A late illustration of the house in 1865 shows the accuracy of this description (Plate 10).[8] In Kirkwood's map of 1817, Black's house is numbered No. 26 and is on the corner of Hill Place.[9] Nicolson Park (or Lady Nicolson's park) was originally the house of Sir James Nicolson of Nicolson and Lasswade, Bart. After his death in 1743, with housing in Edinburgh in great demand, his widow Elizabeth Nicolson built Nicolson Street, the new road out of Edinburgh and Nicolson Square.[10] This became an area of new houses which were in marked contrast to the mediaeval buildings of the Old Town or College Wynd, but not as grand (or seemingly as well built) as the buildings of the New Town slowly rising on the side of the Nor' Loch (now Princes Street Gardens). This location appealed to Black, who was anxious to be close to his laboratory – indeed, so much so that he asked Robert Adam to build a house for the chemistry professor in the new university buildings that Adam was designing. In the event his request was turned down, and in any case the new university buildings (now called Old College) were not erected for more than two decades after Black's death. He had two servants in the house with him, a housekeeper and a manservant. He presumably kept chickens (or hens, as he would have called them) at the back, as his household accounts have entries for the purchase of hensmeal and birdseed.[11] His relatives sometimes stayed with him, including one of his nephews, Jamie Black, who died of tuberculosis there in 1784.

George Black, Joseph's nephew and his executor, sold the house to Margaret Rollo, the widow of Alexander Houston, a banker, in 1800.[12] In 1816 it was bought by a confectioner, James Waddell (also spelt Weddell). It seems to have been used as a shop, as it was shared with another confectioner, William Hill, and Waddell also had a mansion at Pendreich, on Eskbank Road, between Lasswade and Bonnyrigg, some seven miles down the road.[13] As she was a widow, it is quite possible that Margaret Houston had used it as a shop as well – the more so since it seemed to have been legally linked to Alexander Ainslie, merchant. In 1822 Waddell became bankrupt, and the house was taken over by the Edinburgh Savings Bank as one of his creditors. It appears that the Asylum for the Industrious Blind bought it from the bank to set up a female blind asylum (Plate 10).[14] The Edinburgh Asylum for the Relief of the Indigent and Industrious Blind had been founded in 1793, the third blind asylum in the world after Paris and Liverpool, and a nearby house in Nicolson Street (now 58 Nicolson Street) had been purchased for the male blind in 1806 under its new name of the Asylum for the Industrious Blind.[15] Obtaining a large house for the female blind a few yards up the road was thus a logical move. In the Ordnance Survey map of 1849–53, the house of the corner of Hill Place is marked 'Asylum for Industrious Blind (Female)'.

There is an early (c.1820) picture of the Blind Asylum by J. & H.S. Storer which is often reproduced as Black's house, on the assumption that if it was the Blind Asylum it must have been Black's house.[16] However, it cannot be Black's house, because that still belonged to Waddell in 1820. Furthermore, it is a two-storey building, which is nothing like the description of Black's house. Finally, there is clearly a public house next door, which can be seen next to the male blind asylum at 58 Nicolson Street on the Ordnance Survey map of 1877.[17] So clearly this is a picture of the original Blind Asylum at 58 Nicolson Street. The small building in this picture was demolished in the 1830s or 1840s and replaced by a shop selling goods made by the blind people at the street front, and a workshop at the back.[18] Somehow, by the late nineteenth century, the assumption had arisen that Black's house was the male asylum, not the female one. In James Grant, *Edinburgh Old and New*, the reference to Black's house in Nicolson Street is ambiguous,[19] but the picture a few pages later is the one by Storer of the male asylum, and it may well be the source of this error.[20] Similarly, in *Romantic Edinburgh*, John Geddie states that Black's house later became the Royal Blind Asylum (the later name of the Asylum for the Industrious Blind), leaving the reader to come to the erroneous conclusion that it was the male asylum rather than the female one.[21] Neither of these local authors seems to have been aware that Black's house had by this time become a bank. This error continues to be propagated because of the mistaken identification of the Storer print with Black's house.

In 1875, the female blind asylum moved to Craigmillar Park, the current site of the Royal Blind Asylum. The street level of Black's former house was then occupied by a branch of the National Bank of Scotland and the other two shops.[22] The upper floors were used as warehouses. Hence the building on the southern corner of Nicolson Street and Hill Place is marked 'Bank' on the Ordnance Survey Map of 1877. By 1911 it had become a branch of the Edinburgh Savings Bank (which is not related to the Edinburgh Savings Bank that took the house over in 1822, as it was only founded in 1836).[23] The Royal Blind Asylum finally sold the building to Louis Cowan, a china merchant, in 1930. Only three years later, at the height of the Great Depression, the trustees of the now bankrupt Louis Cowan sold it to Alexanders Stores of Kirkintilloch for £2,500, plus £5,000 to cover a bond (Plate 11). Alexanders was a chain of department stores specialising in ironmongery, crockery and hardware (in the pre-computers sense of the term). There were other Alexanders Stores in Glasgow, Ayr, Falkirk, Kilsyth and Bo'ness.[24] It was still open in 1972, but a Compulsory Purchase Order was placed on the building by Edinburgh City Corporation in June 1973 within the framework of the planned ring road for Edinburgh (the Eastern Link Road). The whole of Hill Place and this building were part of the proposed Pleasance Junction North, which would have linked the ring road with Nicolson Street.[25] Alexanders Stores were paid £51,562 in compensation.

While the Compulsory Purchase Order remained in place the building could not be used, and judging by photographs of the time it remained in a rather sorry-looking state until the Eastern Link Road scheme was dropped (Plate 11). It was taken over by Charles McKinlay in 1981, and converted into an Italian restaurant and the Uptown Disco. Sadly a fire broke out on the night of 14 November 1982, said to be have been caused by a cigarette end setting alight a bale of hay (which had been brought in

Plate 1. Herman Boerhaave (1668–1738), professor of chemistry at the University of Leiden. By George White, published by Thomas Bowles senior, c. 1700–25.

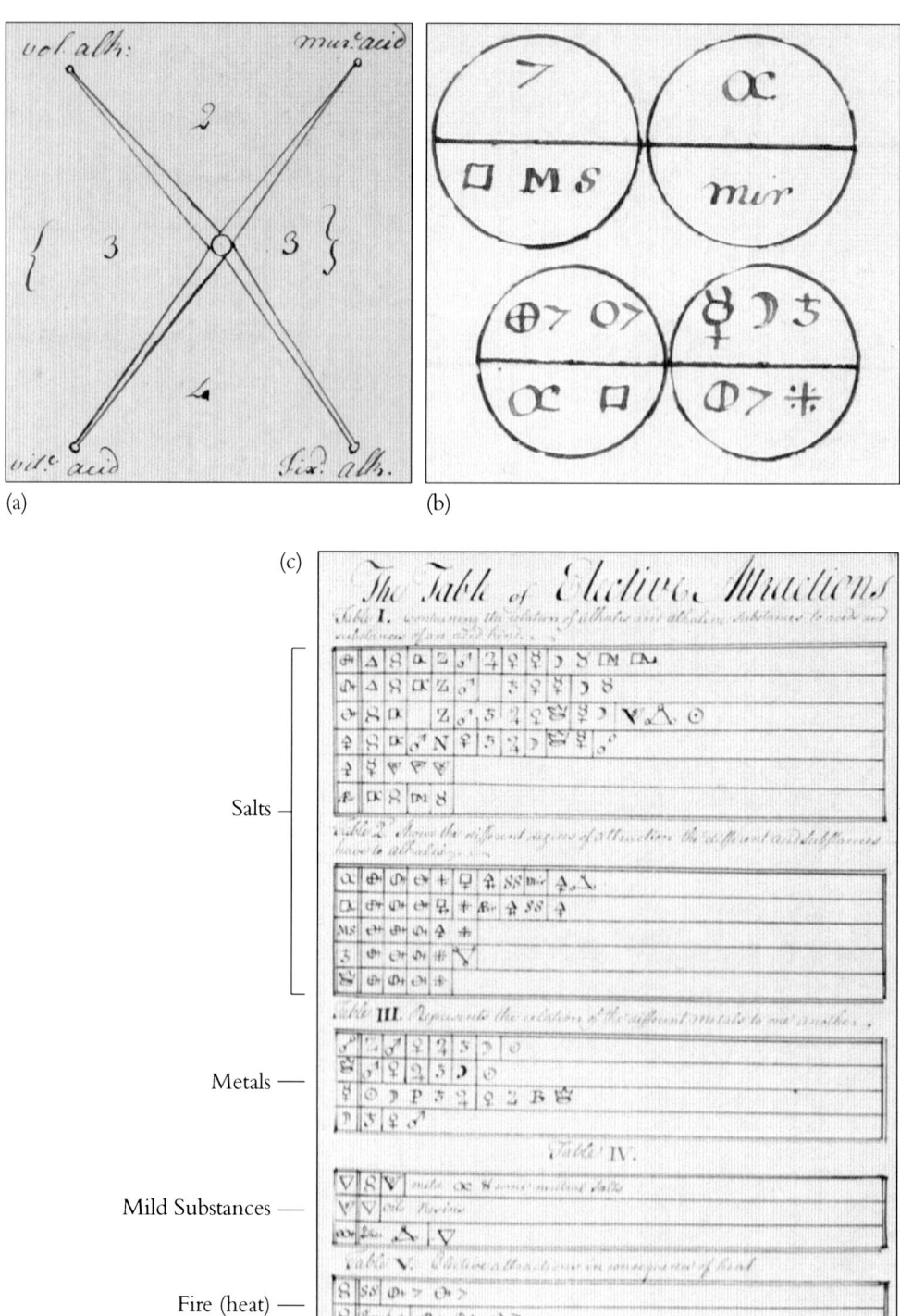

Plate 2. Joseph Black's three main visualisations of affinity: (a) chiasm, (b) circlets, (c) a square affinity table. From Paul Panton's notes taken at Joseph Black's lectures, 1778.

Lect: 61st. —— I shall add another Experiment:
viz to shew yt the air, yt Alkaline Substances actually Contain,
Extinguishes Flame and is Capable of killing breathing Animals. —— For this purpose I put a qty of Chalk & Water
into this Cylindrical Glass Viol, (A) and pour into
it some diluted vitriolic Acid. It produces an Efferves-
cence, separates and Expels yt Air united wt ye Chalk,
throws it off loose and restores it to its Elastic State, & While it
arises from the Chalk, the oyr Common Air, Which ye Vessel
Contains is Gradually driven Out, and after We have added
a Certain qty of yt Acid, the Vessel Will be filled With this
Air, Which is Conveyed by a Tube into a Cylindrical Vessel,
(B) from Which the Atmospherical Air has been expelled
by burning a piece of paper in it. ——

We can't observe that yr is any change produced
in yt upper part of yt Vessel, it is Still transparent, but
upon Examining it a little We shall soon find it to be
differt. from Common Air, for it Will continue in yt Vessel
some time tho' it is Open, yt least Agitation of yt surrou-
nding Air Will Occasion it to be dispersed, upon immers-
ing a Candle into it, the moment yt flame sinks below
the level of yt Vessel, it is extinguished, some times
the flame hovers over the Wick, at a Certain distance
above

Plate 4.

Right: Charles Blagden's unsuccessful attempt to inscribe Black's affinity circlets into his student notebook.

Below: Blagden's reconfiguration of Black's circlets into squares.

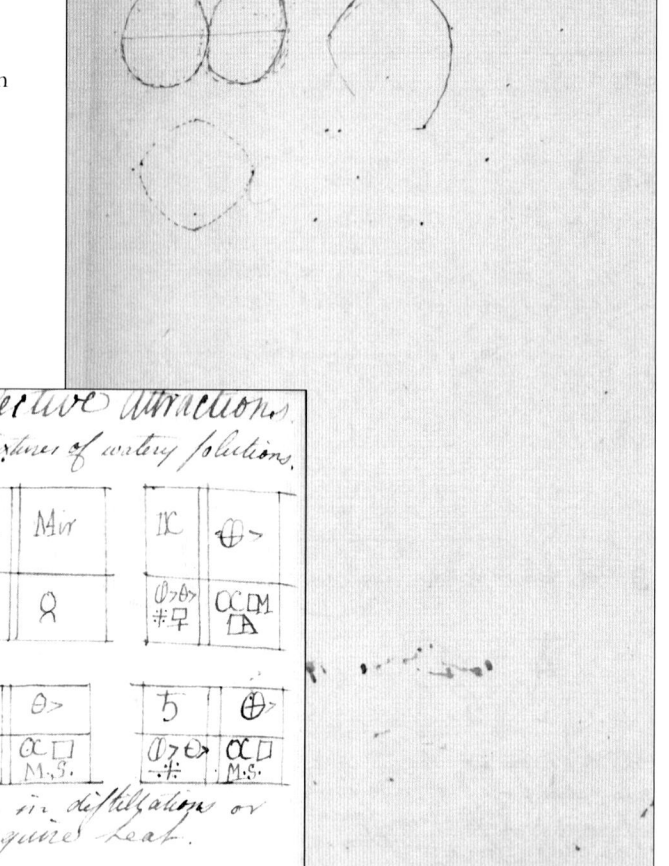

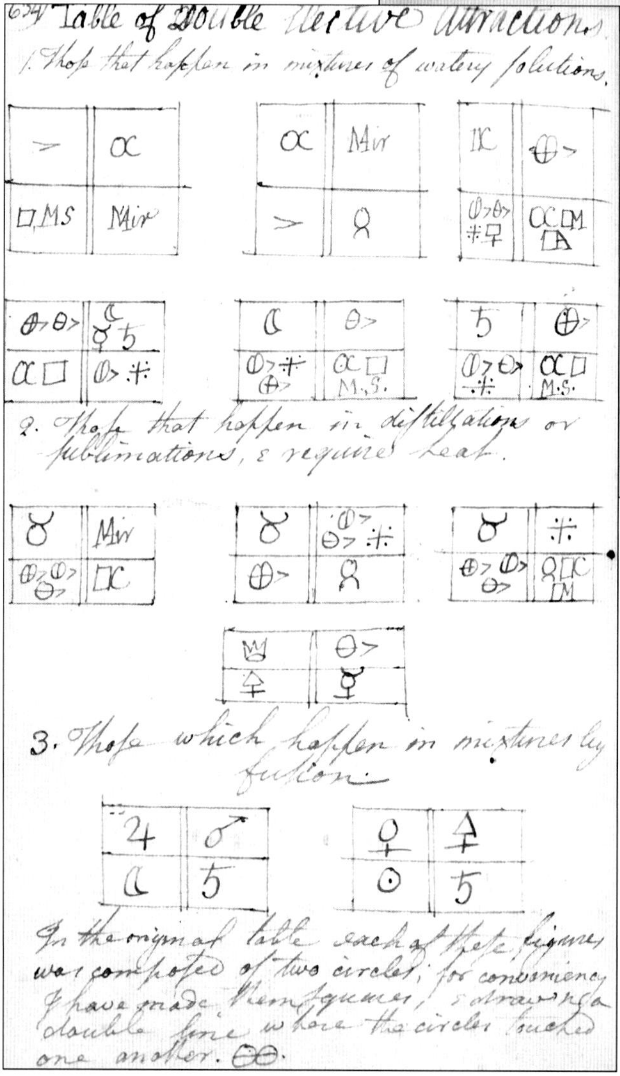

TABLE DES RAPPORTS *Differents observez Entre Diverses Substances Chymiques par Monsieur Geoffroy De l'acad. des Sciences*

Plate 5. *Above*: Venel, Gabriel François (1723–1775), 'Table des Rapports', *Cours de Chymie.*

Plate 6. *Right*: Affinity table in Lewis, William, *The New Dispensatory* (London: Nourse, 1753), p. 11.

A **TABLE** of the relations or affinities observed between different SUBSTANCES.

INFLAMMABLE SPIRITS	Water	Oils and resins					
WATER	Inflammable spirits	Neutral salts, composed of mineral acids and fixt alkalies; and metallic salts					
	fixt alkaline salts	inflammable spirits					
ACIDS in general	fixt alkaline salts	volatile alkaline salts and alkaline earths	metallic substances				
The VITRIOLIC acid	the inflammable principle of bodies	alkalies	zinc	iron	the earth of alum	copper	mercury
The NITROUS acid	zinc	iron	copper	tin, lead	mercury	silver	camphor
The MARINE acid	iron	regulus of antimony	copper	silver	mercury		
FIXED ALKALINE SALTS	the vitriolic acid	the nitrous acid	the marine acid	vegetable acids	oils, sulphur		
VOLATILE ALKALINE SALTS	the vitriolic acid	the nitrous acid	the marine acid				
ALKALINE EARTHS	the vitriolic acid	the nitrous acid	the marine acid				
METALLIC SUBSTANCES	the marine acid	the vitriolic acid	the nitrous acid	vegetable acids	oils		
SULPHUR	fixt salts, quicklime	iron	copper	lead	silver	regulus of antimony	mercury
REGULUS of ANTIMONY	iron	copper					

If the first substance in any of the foregoing series's be combined with another in the same series, the addition of any of the intermediate bodies will disunite them. Thus, if any acid is combined with a metallic substance, it will let go the metal to take up an alkaline earth, or volatile salt; and these again it will forsake, to unite with fixed alkalies. The uses of this table, in many of the capital operations of the present pharmacy, will sufficiently appear hereafter.

CHAP.

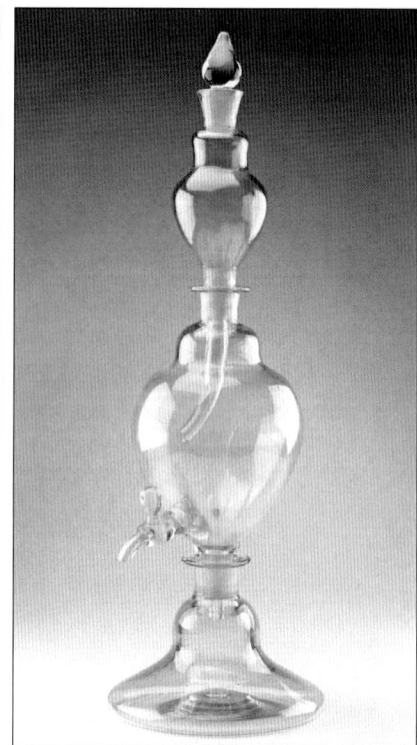

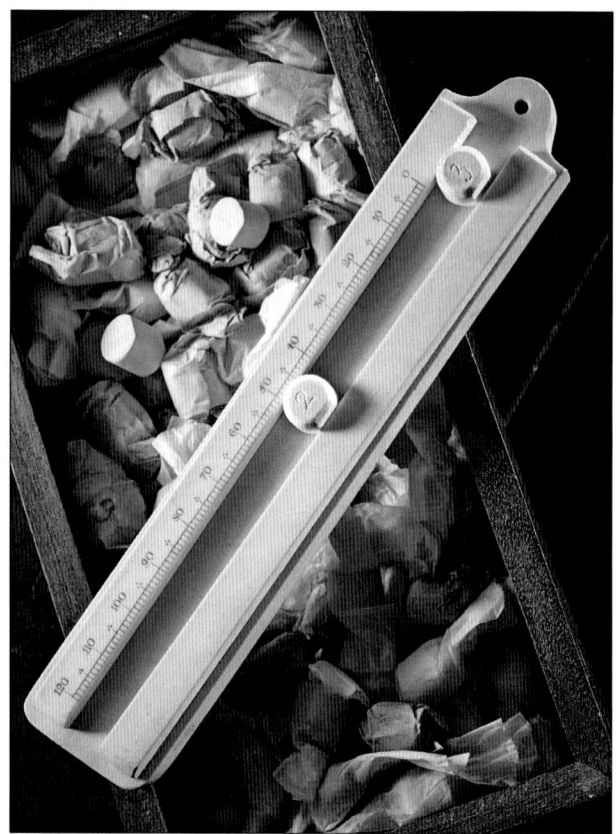

Plate 7. *Above left*: Leith-pattern wine bottle, from near Forres, Moray, late eighteenth century.

Plate 8. *Above right*: Nooth's apparatus for aerated mineral waters, deemed to be good for the health, c.1775.

Plate 9. *Right*: Wedgwood's pyrometer, a temperature-measuring instrument for kilns, first described in 1782.

Plate 10. Female Blind Asylum, Nicolson Street, Edinburgh, in 1865. Formerly Joseph Black's residence.

Plate 11. Alexanders Stores from Nicolson Place, Edinburgh, in 1981.

Plate 12. 'Uptown' disco, Nicolson Street, Edinburgh, after fire, 15 November 1982.

Plate 13. 'Uptown' disco under demolition, 16 November 1982.

Plate 14. Members of the Round Table Club (1877). Joseph Bell is seated, second from the right (arms crossed); Alexander Crum Brown is seated fourth from the right (in the middle, behind the table); Thomas Richard Fraser is standing sixth from left, directly behind Crum Brown.

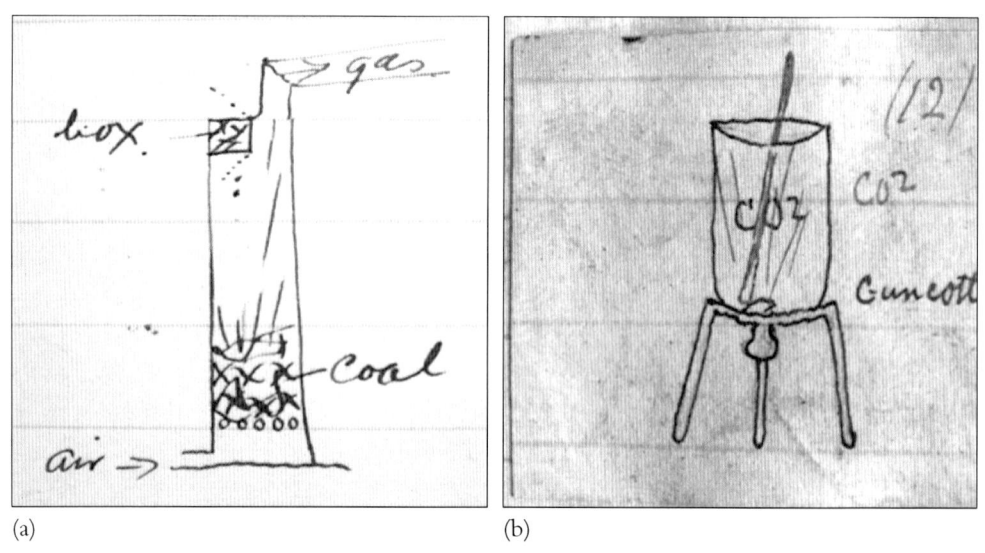

(a)   (b)

Plate 15. (a) Image in John Murray's lecture notes illustrating the demonstration of production of carbon monoxide and carbon dioxide. (b) Image taken from Crum Brown's lecture notebook, demonstrating ignition of gun cotton.

for a Country and Western themed night).[26] The building was completely gutted, and it was demolished the next day on safety grounds (Plates 12 and 13). It was replaced by a new building, which was occupied on the ground level by Kentucky Fried Chicken and McColl's the newsagent in 2013.[27] The new building was briefly owned – rather appropriately – by the nearby Royal College of Surgeons of Edinburgh.[28]

The Houses of the Famous

When the building was demolished, did anyone realise it was Black's final house? The article in the *Edinburgh Evening News* announcing its demolition would suggest not, and there do not seem to have been any protests when it was scheduled for demolition in 1973. At the end, it would not have made any difference, as it was impossible to leave the building standing. This lack of awareness raises the general issue of the survival of houses associated with historical figures and more generally, the knowledge that the house or at least the site was associated with a historical figure. Adam Smith's house, Panmure House off the Canongate on the Royal Mile, is at the time of writing being restored by the Edinburgh Business School.[29] William Cullen's house in Mint Court off the Cowgate (also called Mint Close or South Gray's Close) has long disappeared. David Hume's house on the corner of St Andrew Square and South St David's Street has also been long demolished, but its position has been marked by a plaque.[30] The site of James Hutton's house in St John's Hill, overlooking Salisbury Crags, was marked by a memorial garden and a bronze plaque on a block of Clashach sandstone in 1997 when the street was redeveloped to hold student residences (the plaque has since been stolen).[31] The house itself was demolished at the beginning of the twentieth century.[32]

This suggests that the sites of both Black's house and Cullen's house could be marked by plaques to raise general interest in these chemical giants of the Scottish Enlightenment. Indeed, there is already a plaque to Black on Sylvan House.[33] The memory of famous inhabitants seems to be preserved better – if the internet is any guide – when they lived in the countryside. Cullen is remembered as the resident at Ormiston Hill, a farm at Kirknewton on the southwestern outskirts of Edinburgh near the new town of Livingston, and he is buried in the local kirkyard with a tombstone dating from the mid nineteenth century, partly funded by the Royal College of Physicians.[34] The house itself still exists, but is derelict.[35]

Similarly, the philosopher Dugald Stewart's family home at Catrine in Ayrshire is still associated with him and the farm still exists, although the actual house has been demolished.[36]

Notes and References

1 Morris, P.J.T., 'The House that Black Built', *Chemistry World* 1 (June 2004), p. 88.

2 Anderson, R.G.W. and J. Jones (eds), *The Correspondence of Joseph Black* (Farnham: Ashgate, 2012), vol. 1, pp. 49–51; Grant, J., *Cassell's Old and New Edinburgh: Its History, Its People and Its Places* (London: Cassell & Co., 1885–1887), vol. 4, p. 254, online at edinburghbookshelf.org.uk (last accessed 27 March 2014).

3 Hutton, L., *Literary Landmarks of Edinburgh* (London: James R. Osgood, McIlvaine & Co., 1891), pp. 45–6, online at archive.org (last accessed 27 March 2014).

4 Anderson and Jones, *Correspondence of Joseph Black*, appendix 9, pp. 1457–8 (on p. 1457). Interestingly he valued the house in 1799 at £1,400, as he presumably regarded it as a depreciating asset.

5 To add a personal note, my mother was born in one of these streets (Hermitage Terrace, now Rosevale Terrace) in 1912, just over a century later.

6 Anderson and Jones, *Correspondence of Joseph Black*, appendix 10, pp. 1459–66 (on p. 1461). This volume has 'cooking [?]' before 'house pump'. It is clear from the sasine that the adjective is 'working' not 'cooking'.

7 Register of Sasines, RS27/790, pp. 44–56, National Records of Scotland (on p. 45).

8 It appeared on the cover of the *Report by the Directors of the Edinburgh Asylum for the Indigent and Industrious Blind* for 1866, reproduced in *The Royal Blind Asylum and School: 1793–1993. 200 Years of Service* (Edinburgh: Royal Blind Asylum and School, n.d.), p. 6.

9 Kirkwood, R., *This Plan of the City of Edinburgh and its Environs* (Edinburgh:Kirkwood & Son, 1817), http://maps.nls.uk/joins/416.html (right-hand map) (last accessed 27 March 2014).

10 Grant, *Cassell's Old and New Edinburgh,* vol. 4, pp. 334; Gray, J.G., *South Side Story* (Glasgow: W.F. Knox & Co., 1962) https://sites.google.com/site/southsideheritagegroup/the-south-side-story/lady-nicolson-s-park (last accessed 27 March 2014).

11 Anderson and Jones, *Correspondence of Joseph Black*, appendix 7, pp. 1447–54 (on p. 1447 and p. 1452).

12 Register of Sasines, RS27/790, pp. 44–56, National Records of Scotland.

13 Register of Sasines, RS27/938. pp. 68–85, National Records of Scotland.

14 Search sheets for 36–42 Nicolson Street obtained from the Registrars of Scotland.

15 *The Royal Blind Asylum and School*, pp. 5–6, http://www.royalblind.org/our-organisation/our-history (last accessed 27 March 2014).

16 The print has been reproduced as Black's house in various publications and several websites including *The Royal Blind Asylum and School*, p. 5 (despite the correct picture being on p. 6); Anderson and Jones, *Correspondence of Joseph Black* in Fig. 17 on p. 50, and the Look and Learn History Picture Library, http://www.lookandlearn.com/history-images/M063014/The-Blind-Asylum-Formerly-the-House-of-Dr-Joseph-Black-Nicolson-Street-1820?img=1&search=The+Blind+Asylum&bool=phrase (last accessed 27 March 2014).

17 Ordnance Survey town plan of Edinburgh surveyed 1877, sheet 36, available at http://maps.nls.uk/view/74415674 (last accessed 27 March 2014).

18 As shown by the picture of the larger building in *The Royal Blind Asylum and School*, p 6. Unfortunately the Royal Blind School and Asylum was unable to find the date of the demolition of the old building for me.

19 Grant, *Cassell's Old and New Edinburgh*, vol. 4, pp. 335–6.

20 Ibid., p. 340.

21 Geddie, J., *Romantic Edinburgh* (London: Sands & Co., 1900), p. 185, online at archive.org (last accessed 27 March 2014).

22 The later history of the house is taken from the search sheets for 36–42 Nicolson Street obtained from the Registrars of Scotland.

23 See http://www.lloydsbankinggroup.com/Our-Group/our-heritage/our-history/tsb/edinburgh-savings-bank/ (last accessed 27 March 2014).

24 Entry on the Scottish Archives Network at http://195.153.34.9/catalogue/person.aspx?code=NA19050&st=1& (last accessed 27 March 2014); entry at the Falkirk Community Trust at http://collections.falkirk.gov.uk/search.do;jsessionid=D9D573CFA704F2D9733C0D7CBEB49D9F?id=26541&db=person&view=detail&mode=1 (last accessed 27 March 2014), and elsewhere on the internet.

25 Pers. comm from Brenda Connoboy of Edinburgh City Archives, 16 May 2014. Also see https://maps.google.co.uk/maps/ms?msid=209737842711828293504.0004c9abdc0ce1879dfa5&msa=0&dg=feature (last accessed 20 May 2014).

26 Comment by James Catterson in 'Lost Edinburgh' on Facebook, https://www.facebook.com/lostedinburgh/posts/153238318144145 (last accessed 27 March 2014); 'City Centre Disco Destroyed', *Edinburgh Evening News*, 15 November 1982, p. 1.

27 Personal observation, June 2013.

28 Pers. comm. from Brenda Connoboy, 16 May 2014.

29 See http://www.panmurehouse.org/ (last accessed 27 March 2014).

30 See http://openplaques.org/plaques/10316 (last accessed 27 March 2014).

31 Butcher, N.E., 'The Hutton Memorial Garden at St John's Hill, Edinburgh', *GA: Magazine of the Geologists' Association* 1.1 (2002), pp. 18–19. Also see Butcher, N.E., 'The Hutton Memorial Garden', *The Edinburgh Geologist* 38, http://www.edinburghgeolsoc.org/edingeologist/z_38_06.html (last accessed 9 May 2014).

32 Butcher, N.E., 'James Hutton's House at St John's Hill, Edinburgh', *Book of the Old Edinburgh Club*, n.s. 4 (1997), pp. 107–12.

33 See http://www.geograph.org.uk/photo/3258090 (last accessed 27 March 2014).

34 See http://www.cullenproject.ac.uk/william-cullen.php (last accessed 27 March 2014).

35 See http://www.britishlistedbuildings.co.uk/sc-13646–ormiston-hill-house-kirknewton (last accessed 27 March 2014).

36 See http://www.ayrshirehistory.com/catrine_nether_catrine.html and http://catrineayrshirecouk.ipage.com/cathouse.htm (last accessed 27 March 2014).

TEN

Thomas Charles Hope and the Limiting Legacy of Joseph Black

ROBERT G.W. ANDERSON

In an apologia compiled towards the end of his life, Thomas Charles Hope, the fifth professor of chemistry and medicine at Edinburgh, wrote the following:

> Those who devote themselves to the science of chemistry, may be divided into two classes – 1st, Those whose labours are employed in original researches, to extend our knowledge of the facts and principles of science. 2ndly, Of those whose business it is, from university or other appointments, to collect the knowledge of all that has been discovered, or is going forward in the science, to digest and arrange that knowledge into lectures, to contrive appropriate and illustrative experiments, and devise suitable apparatus for the purpose of communicating a knowledge of chemistry to the rising generation, or others who may desire to obtain it. From my professional situation, I consider myself, as Dr BLACK has done before me, as belonging to the second class of chemists. I consider my vocation to be the teaching the science.[1]

This is revealing for two reasons: first, Hope, who taught until 1843, saw himself as continuing in Black's tradition, even 40 years after his mentor's death. Secondly, that as a consequence, he had not been persuaded to adjust his approach by the very significant advances which had been made in the development of research training laboratories in Germany and elsewhere. This development had been occurring since the beginning of the nineteenth century and was certainly known to Hope, whose contemporary Thomas Thomson set up a teaching laboratory in Glasgow as early as 1818.[2] This paper asks why Hope took these approaches, and considers his career in relation to that of Black. Another question to be considered is whether Hope's enduring reputation as a conservative, pompous, rather dull teacher is justified.

Black was born into a mercantile family, based at Bordeaux, no member of which had worked in a profession or had been to university. This is in sharp contrast to Thomas Charles Hope, who was born in Edinburgh in 1766 and whose father, John Hope, was professor of botany at Edinburgh. The family was Establishment, through and through. His great-grandfather was Sir Archibald Hope, Lord Rankeillor, a Lord of the Judiciary. Two of Hope's brothers were lawyers and a third was an army officer. By contrast, Black's brothers were largely unsuccessful Ulster merchants who got into the habit of going bankrupt at regular intervals.

However, Black and Hope had not dissimilar educations. Black went to the Latin School in Belfast and Hope to the High School of Edinburgh. Black matriculated at Glasgow University at the age of fourteen, while Hope went to Edinburgh University even younger – he was only thirteen. Both studied the general arts course before turning to medicine. Black's inspiration was William Cullen, while Hope's was Black himself. Both graduated at Edinburgh, because Black transferred there from Glasgow halfway through his medical degree course. Black received his MD in 1754 while Hope gained his in 1787.

A year before he graduated, immediately after his father's death, Hope made efforts to be appointed professor of botany in his father's place. He had strong support, particularly from Sir Joseph Banks, President of the Royal Society of London. But it was the aristocratic powers behind Scottish academic thrones which usually prevailed at these times,[3] and the job went to Daniel Rutherford, discoverer of nitrogen, and son of John Rutherford, one of the first four medical professors in Edinburgh.

Black was at a bit of a loss after he graduated – he wanted to make a career in chemistry but was uncertain about how to proceed. Andrew Plummer's disabling stroke a year later, in 1755, made the Edinburgh chair possibly available to him and he had strong academic backing for it – but in the end, it was William Cullen who was appointed. Both Black and Hope started their careers by being appointed to the same Glasgow chemistry lectureship, although they occupied it 31 years apart: Black took over Cullen's job in 1756, and Hope succeeded William Irvine in 1787.

Black's doctoral dissertation was much more distinguished than was Hope's. Black's work on alkalis ultimately led to the identification of carbon dioxide, and it was an inspired piece of work. Hope's thesis, titled (in translation from Latin) 'On the Movement and Life of Plants', was relatively mundane. Black was averse to travel and though he briefly entertained the possibility of going to London after graduation, he changed his mind, and it would be another 34 years before he got to the British capital (on what might have been his sole visit). Hope travelled relatively little, too, but he did manage an important visit to Paris when he was still a young man.

The post-graduation activities of Hope have been well rehearsed elsewhere, though as background, the journeyings of Sir James Hall of Dunglass, the gentleman geologist and friend of the Hope family, should be considered. He departed from Edinburgh on his Grand Tour on 9 May 1783.[4] Reaching Rome at the end of the following year and hearing that Vesuvius was about to erupt (a report which, unfortunately for him, turned out to be false), he dashed to Naples in March 1785. On his way home he stopped off in Paris, in February 1786, and met the outstanding chemist of the day, Antoine-Laurent Lavoisier, from whom he gained a standing invitation to dine on Mondays. The oxygen theory was on its way to being established, though a number of eminent figures of the chemical world were to hold out against it for many years. Hall was convinced of its veracity, partly by his own intuition and partly by his uncle's convictions, and on his return to Edinburgh he presented two papers early in 1788 on the new theory of chemistry at the Royal Society of Edinburgh. Sadly, little is known about what was said, as they were not published in the Society's *Transactions*. Hope decided to spend his 1788 summer vacation in Paris, and there wrote a letter which records that he was well received 'by M. Lavoisier & his set'. It is likely that it was Hope, in Glasgow, who first transmitted Lavoisier's ideas to a student audience in Britain. In Edinburgh, students took up this chemical revolution more enthusiastically than did their professor, and they were debating the theory favourably as early as 1786. Black was famously diffident about any successor to the phlogiston theory. It is true that he mentioned the new French ideas to students in his 1783–4 lectures, but only to dismiss them. His final capitulation did not occur until October 1790, in a now-famous letter to Lavoisier.[5]

With a job under his belt, Hope's research started promisingly, as had Black's. Earliest, and perhaps best known, was his work on strontium salts. A new mineral had been recognised which came from a quarry near the remote village of Strontian in the Western Highlands. Hope performed a series of experiments showing that it was a 'hitherto unknown earth'. He reported this to the Royal Society of Edinburgh in November 1793, but unfortunately for Hope this work was not published in its *Transactions* until 1798.[6] In Berlin, Martin Klaproth was independently conducting similar experiments, and managed to get them into print in September 1793. Others, too, were also working on strontium salts, so Hope can only be called one of the discoverers of the new element. The metal itself was not isolated until 1808 when that great Royal Institution electrolyser, Humphry Davy, produced a sample.

Hope's other researches divide into those on heat and those on plant natural products. In the former, he could be said to be carrying on the Black tradition; in the latter, he was following in his father's footsteps. In between these two bursts of activity lay a great gap when nothing was published. In 1804, Hope read a paper to the Royal Society of Edinburgh on the anomalous change in the density of water with changes in temperature, and the apparatus he devised became a common feature of laboratory demonstrations. In his published paper of 1805, he also addressed the question of whether liquids could conduct heat. Rumford had said they could not; Hope showed that they could.[7] Thirty-four years later, Hope addressed the question of the temperature at which sea water attains its maximum density,[8] and this signalled a minor burst of research activity. Around that time he also published papers on the colouring matter of flowers and on chemical nomenclature. In his final years he read a further paper on the action of reagents on infusions of flowers, and on an anomalous climatological phenomenon, the Freezing Cavern of Orenburg.[9] None of these added up to much, and for 30 years, in mid-career, he made no contribution at all. His judgement on himself – that he did not employ himself in original researches to extend the facts and principles of chemistry – does indicate candid self-knowledge, and possibly reveals embarrassment.

It has become well known that Black, although he essentially abandoned research in philosophical chemistry after his 1760s work on latent and specific heats, did not give up chemical experimentation. He threw himself into industrial consultancy and acted to promote the development of chemically-based industrialisation by providing a great deal of advice to others. This has become clear with the recent publication of Black's complete surviving correspondence.[10] Hope appears to have done nothing of this kind of work, and it is not unreasonable to ask how Hope employed his time between the end of his courses in May and the beginning of the next university session in November.[11] The only activity which slightly approaches Black's huge enthusiasm for advising others is a report which Hope compiled in 1813 with Thomas Telford to test the Edinburgh water supply.[12]

Having said that, there is no doubt that Hope was extremely conscientious in providing his medical students with a systematic and rigorous course of chemistry every year. It was not just medical students who signed on for the courses, and this is clear on examining the registers which list students who paid their fees to attend.[13] In the first year following Black's death, 1799–1800, numbers were 293. These rose gradually up to a peak of 559 in 1823–4, after which they declined to 118 in 1842–3, the last session at which Hope taught. Traill estimated that Black taught chemistry to 15,500 students through the course of his life, though only a small minority of these would become medical graduates. The lecture room was at times so full that some students had to sit on the floor. This evidence is found in Hope's manuscript notes in a packet labelled 'Electricity Old Notes Principally', where he wrote 'When an electric shock is given to students – Make a long train by those on the floor to the right join hands with those in the front bench & those at East End of the bench join with those on floor to the left.'[14]

When he felt his powers to be waning, Black was instrumental in bringing Hope over from Glasgow in 1795 and having him nominated his designated successor in favour of Black's assistant from 1791, John Rotheram. During the 1796–7 session, the two men taught alternate blocks of lectures. This was the last time that Black taught.[15] Hope continued very closely in Black's footsteps, as student lecture notes for the session 1800–1, now in the Chemical Heritage Foundation in Philadelphia, testify.[16] The length of the course was identical to Black's, starting on 29 October and concluding on 30 April the next year: a total of 137 lectures. The writer of these notes, Bertie Greatheed, recorded that Hope had performed 462 experiments in front of his class during the session. The closeness of the approach to Black's own course is further emphasised by stuck-in printed tables showing a list of pharmaceutical preparations of mercury and antimony. These were originally produced by Black for his own courses. There are also close similarities to another set of notes of 1809–10 written down by, or for, Joseph Walker.[17] Though it is difficult to be quite certain how many subject heads Hope spoke to, there are approximately 120 in this set, starting with Attraction and finishing with Urine. A third set of 1826–7 appear to be written by a much more engaged student, who writes in a more informal style and peppers his pages with sketches of the experiments which were being performed in front of him. In one section, he wrote in his notes; 'Missed lecture through mistake I shall therefore sketch Dr Hope's instruments stuck by the fireplace' and an accompanying sketch shows hoops, spouts, rings, a deflagrating ring, a syphon, etc. This is a rare insight into the contents of the laboratory. In a third set, a glimpse can be caught of the demonstrations; for example, samples of minerals were brought into the theatre on trays to show to students, the notes containing sketches of these. Another set of student's notes makes reference to this section of the course, where he writes a note: 'X. During the whole of these discussions on the Earth & Fossils the most splendid Collection of Fossils was exhibited & illustrated with all that Precision & Acuteness which is peculier to Dr Hope.'[18]

At the very end of the course, Hope signed off in a somewhat self-satisfied manner, as follows:

I have now gentlemen gone through the plan of my course and the period has now arrived to bring these lectures to a conclusion.

I am willing to satisfy myself that you are now acquainted with the extent of the subject and I trust that you have derived much usefull information – I now lay before you the method of pursuing the subject and you are now able to peruse the larger systems of chemistry as Thomson & Murray. And you see chemistry is not to be obtained in the closet but by experiment and Dr Henries[19] is the best work. May I be indulged in the hope that many of you may be stimulated with the hope of presenting the science of chemistry. There's not a more extensive field of discovery or that will more reward the fruitfull skill of the discoverer than Chemistry the richness of this country has grately benefitted within this last century. Every day discloses something new. The art of experiment is not to be obtained without much difficulty and nicety of attention; after these remarks may I be permitted to conclude wishing you all success and all happiness.

Hope then writes ' – FINIS'.

It would be wrong, which has been suggested, that Hope's lectures were changeless and became out of date. Inspection of his own notes, from which he taught, preserved in Edinburgh University Library, bears this out.[20] Each lecture is written out on small pieces of stiff paper, kept together in stout envelopes, one per lecture. Hope's notes were constantly being revised, with additions sometimes interwoven into his text, sometimes written on separate pieces of paper and pasted over the earlier notes. They indicate a tidy mind, and contrast strongly with Black's notes, which John Robison nearly despaired of, saying that 'Many of them are on scraps of paper, much altered and interleaved' and 'many scraps were only memorandums'.[21] A number of Hope's envelopes contain printed cuttings from articles in periodicals. As examples, there are 1841 cuttings from the *Literary Gazette* and *Philosophical Magazine* which relate to Fox Talbot's development of the Calotype process, and an article on electrotypes from the *Morning Herald*. In the business of keeping up with new developments in his subject, Hope sometimes excised sections completely; on one envelope he wrote: 'Old & Obsolete N.B. they would not be the worse for a little Conflagration'.

This process of revision went on continually until the very last lecture of the session which Hope delivered in mid 1843. He wrote himself notes summarising his performance at the end of each session, and after this one, which he did not know would be his last, he noted: 'Progress 1842–3 I was very much rushed in the latter part of this course & had to abridge valuable matter, both of veg. & specially animal matter Ergo in 1843 push vigorously from the start – Specially abridge on Heat, which occupies too much of the Course'.[22]

It is likely that in some areas of chemistry he was more up to date than others. Dalton's atomic theory, disseminated around 1807–8, was taught in Hope's class by 1809. One of his students recorded:

Mr Dalton maintains that when bodies unite in the proportion they combine in proportion to the Particles of which they are respectively formed, and that they vary in the ration of 1:2:3:4 &c. but never as 2:3:4 or so on. Thus he estimated the relative weights of the particles of which the different bodies are composed and this he has done with the integrant Particles of a vast number of bodies.[23]

In one sketch, Hope attempts to explain the different nitrogen oxides diagrammatically. On the other hand, the organic chemistry section was scarcely more than an afterthought. In a note to himself of 29 April 1837, Hope wrote 'By pushing on at the Beginning of the course I got fully as much time as usual for Organnic [sic] Chem^y – (Yet I should have wished to have at least one Lecture More) [. . .].'

The content of these early 'organic' lectures are of some interest. They came under the heading 'Of Animal Matters' – and Hope preceded the first lecture with 'Our chemical knowledge on the Subject is still less Advanced than on Vegetables. Some progress however has been made, which is daily Advancing –'. There then follows a series of lectures on Bones, Flesh, Fluids, Blood, Chyle . . . and so on down to Urine, seven lectures later. The approach was obviously biased by having to provide for his medical audience. It is worth pointing out that Hope's name does not appear in the extensive 1894 second edition of Carl Schorlemmer's *The Rise and Development of Organic Chemistry*.

How might Hope's lectures be judged? They can be considered from the point of view of content and of style. Content-wise they seem fairly up-to-date where providing dates is possible, and it is clear that Hope intended to provide a full course of chemistry, as evinced by his and his pupils' lecture notes. There are, however, blind spots, as has just been mentioned. His colleague Thomas Traill, who had stood in for Hope in the 1843–4 session after Hope had suddenly resigned his chemistry chair, wrote in his memoir of Hope:

> I found his lectures far more nearly written up to the advanced state of chemistry at that period, than I had been led to expect; and although it was necessary to make various alterations and additions, especially in the disquisitions on organic chemistry, these alterations and additions were far less extensive than I had anticipated.[24]

Secondly, there is his lecturing style, which is more difficult to judge, and here there is contradictory evidence. Traill says that Hope's lecturing style was 'methodical and clear, though his style was occasionally too laboured'. The American chemist Benjamin Silliman wrote that students complained of his 'chilling, unsympathizing manner', while another student had reported that 'his manner and diction were pompous'. Several, however, thought that his presentation was clear, and his skill in conducting experiments in front of his large class was widely appreciated. Charles Darwin, who attended medical classes in Edinburgh between 1825 and 1827, wrote: 'The instruction at Edinburgh was altogether by lectures, and these were intolerably dull, with the exception of those on chemistry by Hope.' Silliman compared him favourably with Black, saying 'I was forcibly struck with the great resemblance of Dr Hope's lectures, in style, substance, and illustrations, to those of his great master.'[25] This, combined with the fact that on seven occasions more than 500 students registered for Hope's annual course of lectures – most of whom did not need to attend – scarcely reflects an inadequate, unpopular teacher.

There are two other aspects of Hope's teaching which need to be examined – his popular courses in chemistry, and his attitude towards establishing courses in practical chemistry. It is clear, though little evidence is available, that Black established a special chemistry course for lawyers. James Boswell attended this in 1775, saying that the course bored him, apart from Black's exposition on gold. One of Black's pupils complained that by the time he got to teach his regular class at 11 a.m., Black was dead tired. The idea of public lectures to specialist audiences in Scotland, if that is not a contradiction, may have sprung from William Cullen's lectures on chemistry, vegetation and agriculture of 1768. In any case, Black's course was certainly early of its kind, even in comparison with activity in London. Little is known of Hope's lecture course of 1800, requested by a number of lawyers from the Faculty of Advocates. It is worth pointing out that the first courses of lectures at the Royal Institution in London started in March of that year, which comprised a fashionable course in the afternoons and a 'full and scientific course of experimental philosophy on the plan generally adopted in universities' in the evenings (these universities cannot have been either Oxford or Cambridge, which did not consistently offer such courses at the time). It is not known which type Hope's more closely resembled. There is much greater clarity about the two courses given in 1826 and 1828 in the extravagant new suite of laboratories, which had been built in the long-delayed Old College (called at the time 'New College', naturally).[26]

Hope's 'Popular Lectures on Chemistry' commenced in February 1826. Twenty-two lectures, each lasting for one-and-a-half hours, were delivered three times a week until April, and they were designed to cover 'The General Principles and more important facts of CHEMISTRY'. He demonstrated some of his more spectacular experiments to audiences which included women. Unfortunately we have no visual representation of these classes, as we do with the widely-reproduced Gillray caricature of 1802, 'Scientific Researches!', showing a demonstration going terribly wrong at the Royal Institution (there are various interpretations of this). The nearest image of a chemistry lecture in Scotland (and then not very close) is a print showing Andrew Ure lecturing at Anderson's Institution in Glasgow in 1825.[27] Hope's popular lectures were maliciously reported by the lawyer and member of the Edinburgh literati Henry Cockburn, who wrote: 'the ladies declared that there never was anything so delightful as these chemical flirtations. The Doctor is in absolute extasy with his audience of veils and feathers, and can't leave the Affinities.'[28] The popular course's content, defined in Hope's carefully preserved notes, was a fairly rigorous one, covering heat, pneumatic chemistry, acids, salts and electricity. The subscription was two guineas per person and Hope offered the proceeds of his courses to establish prizes for his students of chemistry.[29] This amounted to the significant sum of £800, or perhaps £17,000 in today's money. Clearly these exercises in popularisation were successful, and the question arises of why Hope did not repeat his public course after 1828.

Hope signed off the 1826 and 1828 courses with the following comments: 'You are now aware that Chemistry is an Experimental Science & will readily be convinced that it cannot be pursued by study in the closet alone – It is necessary to engage in the labors of the Laboratory – Experiments must be performed & processes conducted.' This may seem to be a surprising valediction, considering what we know about his cool attitude to demands for practical courses from his regular students. His former colleague Robert Christison wrote of Hope in 1814: 'Neither did he encourage experimental inquiry among his students. His laboratory was open to no one but his class assistant, Dr Fyfe [...] There was at that time no opportunity for students to learn history practically, either at the great chemical school of Edinburgh, or, indeed, anywhere else in the United Kingdom.'[30] That there was a demand is undoubted, as Christison continues: 'Syme, my twin-brother and I, Andrew Coventry, Alexander Jackson, Robert Mercer, James Hogg, and five others, therefore formed for the purpose a Chemical Society, which met once a-week in the evening and performed and demonstrated such of Dr Hope's experiments as were within our means.'

Hope seems to have had a totally confused attitude towards the provision of laboratory facilities, an attitude which hardened as he got

older and resulted in his denying the need for practical experimentation by students.[31] In April 1823, he occupied his new extensive suite of
rooms in the south-west corner of the Adam-Playfair quadrangle, which
comprised a classroom for 500 students, a separate laboratory, a museum
for the display of minerals and chemicals, a private sitting room, and
three storage rooms for apparatus and materials.[32] Hope wrote to the
University Senate to announce that his assistant, John Anderson, would
be giving an introduction to practical chemistry and pharmacy in the
new building so that students could become familiar with the operations
of chemistry. Anderson was unreasonably restricted by Hope, however.
The laboratory was open only for one-hour sessions three times a day, and
Anderson had to supply all apparatus and materials himself. Needless
to say, Hope creamed off a proportion of the income. These arrangements remained in place for five years. In 1828, the situation changed
when an ambitious extramural lecturer called David Boswell Reid took
over the practical class and his lectures became well attended. When
the Royal College of Surgeons announced that Reid, having no MD
degree, was unqualified to teach, Hope's subterfuge was to deliver the
first class himself, then leave the rest of the course to Reid. Subsequently,
Reid decided that it was in his best interests to become medically quali-
fied, and he undertook the necessary medical classes, graduating in 1830.
Hope's rapaciousness got Reid down, and in 1833 he petitioned the Town
Council, who were patrons of the University, to establish an independent
chair, or at least a lectureship, of practical chemistry. This was bitterly
opposed by Hope, who foresaw that both his income and status would
decline. It was Hope's influence that won the day.

Thereafter Reid decided to part company with Hope, and he offered
complete private courses in chemistry in Edinburgh until 1847, by which
time he had transmogrified into a ventilation engineer.[33] Hope failed to
find another assistant to teach the subject, and with his advancing years
(and 'love of ease', according to Traill), the teaching of practical chemistry
in the university was dropped. It can be argued that this was the most
significant factor in the decline of chemistry as a subject in Edinburgh.
It is somewhat ironic that in the late 1740s Cullen complained that the
Edinburgh chemistry course was too closely confined to the needs of
medicine and that it needed to be broadened,[34] which indeed it was
under Cullen himself and Black. And yet, a century later, the decline
of chemistry can be explained by its again being too closely allied to

the needs of the medical faculty and the training of doctors. It is telling to discover which, and how many, medical students made their way to Giessen to study with Liebig after graduation. Perhaps most significant was William Gregory, who made the journey there in both 1835 and 1841. He is specially mentioned here because it was to be Gregory who would succeed Hope in the Edinburgh chair. Other significant Edinburgh MDs who made the pilgrimage were Thomas Anderson, Robert Harley and William Henry.[35]

In conclusion, it would be useful to provide an overall assessment of Thomas Charles Hope as chemistry professor at Edinburgh. It is important to look beyond the casual remarks which were made about him by commentators such as Henry Cockburn. It can be amusing to be critical, and Cockburn had his own reputation to live up to. Some held a general view that the medical faculty had declined as soon as the New College was complete. As the iconoclastic professor of comparative anatomy at London, Robert Edmund Grant, put it in 1833:

> The University of Edinburgh so long the Athenaeum of Europe when in its glory, in the days of the Blacks, the Cullens, the Monros the Gregories, the Stewarts, and the Playfairs, consisted [...] chiefly of an unseemly aggregate of ancient cottages, where the light of genius shone brightest in obscurity, and unadorned, and in proportion as its place has risen, that University has sunk.[36]

The undeniable fact is that Hope's courses were very popular, even more so than Black's in terms of annual subscription levels. It is true that they tailed off towards the end of his career, but this may be due to two reasons: firstly, there was ever-increasing competition from extra-mural teachers, who undermined the cost of the college-based courses. Secondly, the public fashion for learning about chemistry declined towards the middle of the century. This is confirmed by the Edinburgh statistics. In 1799, when the chemistry class attracted 293 subscribers, there were 52 medical graduations, or 18% of the class. For the huge class of 1823, which attracted 559 subscribers, the 93 graduands comprised 17% of the total. In 1836, the figures are 182 subscribers, 123 graduates being 68%, or two-thirds; while in Hope's last year, there were 87 graduations, the modest class being only 118 strong, with the percentage of medical graduates being 74%, or three-quarters of the class. Thus the number of medical graduations certainly fluctuated over this period, but nothing like to the extent of the total class size.[37]

Hope was undoubtedly conscientious – his continual attention to refreshing the notes from which he lectured bears this out. He was frequently praised for the success of his experimental demonstrations. It is possible to give a perfectly good quality course of lectures and at the same time to be dull, and it has to be accepted that Hope's manner of delivery was not inspiring. He was a solid Tory, often opposed to change and rarely innovative. This is borne out by the way he dragged his feet over the introduction of practical courses, and also by his opposition to the introduction of a new chair of comparative anatomy in 1817. There is a splendid caricature by John Kay, titled 'The Craft in Danger', showing a number of professors attempting to stop John Barclay riding in to the Old College on the skeleton of an elephant. Hope is seen as one of them, rope in hand, jamming his foot on a rock labelled STRONTIAN. A caption above reads ' – Hope is lost the rope gives way/and muscular motion wins the day'.[38] He was certainly conservative in his opinions and approaches, and the caption sums him up well. It can be said that Hope was good for the training of doctors, but that he missed the opportunity to promote his subject in the way that others were doing elsewhere in Europe. Edinburgh would drop several places on the chemistry-teaching league table in the first half of the nineteenth century. No really significant chemist was to emerge from Hope's chemistry programme.

It can be said that his successors scored little better – Gregory, Playfair (surely a particular disappointment, given expectations) and Crum Brown. It was University College London which would take the initiative, following its foundation in 1826, while the Royal College of Chemistry, London, from 1845 (where Wilhelm Hoffman taught) must not be forgotten. Ironically, UCL's chemistry department was led by one Scots-chemist-trained-in-Germany after the other, right up to the beginning of the First World War.[39]

Notes and References

1   Traill, T.S., 'Memoir of Thomas Charles Hope', *Transactions of the Royal Society of Edinburgh* 16 (1849), pp. 419–34 (on p. 431). A brief biography of Hope is Doyle, W.P., *Scottish Men of Science: Thomas Charles Hope M.D., F.R.S.E., F.R.S. (1766–1844)* (Edinburgh: University of Edinburgh History of Medicine and Science Unit, 1982). For further detail, see Anderson,

R.G.W., *The Playfair Collection and the Teaching of Chemistry at the University of Edinburgh, 1713–1858* (Edinburgh: Royal Scottish Museum, 1978), pp. 36–45.

2 Morrell, Jack, 'Thomas Thomson: Professor of Chemistry and University Reformer', *British Journal for the History of Science* 4 (1969), pp. 245–65 (on p. 253). Other useful references to Hope can be found in Morrell, Jack, *Science, Culture and Politics in Britain, 1750–1870* (Aldershot: Ashgate, 1997).

3 Emerson, Roger L., *Academic Patronage in the Scottish Enlightenment: Glasgow, Edinburgh and St Andrews Universities* (Edinburgh: Edinburgh University Press, 2008).

4 Ingamells, John, *A Dictionary of British and Irish Travellers in Italy, 1700–1800* (New Haven, CT: Yale University Press), pp. 443–5.

5 Anderson, Robert G.W. and Jean Jones (eds), *The Correspondence of Joseph Black* (Aldershot: Ashgate, 2012) vol. 1, p. 47; vol. 2, pp. 1110–13 (letter, Black to Lavoisier, 24 October 1790).

6 Hope, Thomas Charles, 'Account of a Mineral from Strontian, and of a Peculiar Species of Earth which it Contains', *Transactions of the Royal Society of Edinburgh* 4 (1798), pp. 3–39.

7 Hope, Thomas Charles, 'Experiments and Observations upon the Contraction of Water by Heat at Low Temperatures', *Transactions of the Royal Society of Edinburgh* 5 (1805), pp. 379–405.

8 Hope, Thomas Charles, 'Inquiry whether Sea Water has its Maximum Density a Few Degrees above its Freezing Point', *Transactions of the Royal Society of Edinburgh* 14 (1839), pp. 242–52.

9 References to these late papers can be found in Anderson, *Playfair Collection*, pp. 36–45.

10 Anderson and Jones, *Correspondence of Joseph Black*.

11 It has to be said that the evidence of Black's industrial interests comes from his fairly extensive correspondence and that relatively little of Hope's survives.

12 Telford, Thomas and Thomas Charles Hope, *Report on the Means of Improving the Supply of Water to the City of Edinburgh* (Edinburgh, 1813).

13 Morrell, Jack, 'Practical Chemistry in the University of Edinburgh, 1799–1843' *Ambix* 16, pp. 66–80 (on p. 76, n. 84).

14 Edinburgh University Library, Centre for Research Collections (EUL), MS Gen 269, packet 50.

15 See EUL MS Gen 48D, 'Notes of Lectures on Chemistry delivered by J. Black and T.C. Hope at Edinburgh University taken down by a student, 1796–97'; also see Doyle, *Thomas Charles Hope*.

16 Chemical Heritage Foundation, Philadelphia (CHF) MS QD14. H674.1800.

17 CHF MS QD14.H674 1809.

18 EUL MS Gen 1399 (vol. III) f. 88v. These notes are of the session 1809–10.

19 Henry, William, *Elements of Experimental Chemistry* (London, 1799); the work went through eleven editions in 30 years.

20 EUL MSS Gen 268–72.

21 Robinson, Eric and Douglas McKie (eds), *Partners in Science: Letters of James Watt and Joseph Black* (London: Constable, 1970), p. 323.

22 EUL MS Gen 270.

23 EUL MS Gen 1398 (vol. II) f.9.

24 Traill, 'Memoir', p. 433.

25 Fisher, George P., *Life of Benjamin Silliman M.D., LL.D.* (London, 1866), vol. 1, p. 165. The Swiss geologist Necker de Saussure, who attended Hope's classes, paid tribute to the nearly magical experiments; see Fraser, Andrew G, *The Building of the Old College: Adam, Playfair and the University of Edinburgh* (Edinburgh: Edinburgh University Press, 1989), p. 193.

26 The Adam/Playfair 'New College of Edinburgh' became officially known as Old College in 1920.

27 Reproduced in *Northern Looking Glass*, 14 November 1825, entitled 'Andersonian Institution'.

28 Cockburn, Henry, *Letters Chiefly Connected with the Affairs of Scotland, from Henry Cockburn to Thomas Francis Kennedy* (London: William Ridgway, 1874), p. 137.

29 In 1837, Hope said to his class: 'Though one individual only can be successful in obtaining the Medal, but every competitor gains a prize of much higher value. He acquires that improvement in his knowledge in the science & in his ability for experimental research, which is my object to encourage', EUL MS Gen 270, envelope 113. If he truly meant this, why didn't he establish a research school?

30 *The Life of Sir Robert Christison, Bart, edited by his Sons* (Edinburgh, 1885), vol. 1, p. 58.

31 Morrell, Jack B., 'Science and Scottish University Reform: Edinburgh in 1826', *British Journal for the History of Science* 6 (1972), pp. 39–56.

32 Fraser, *The Building of the Old College*, pp. 114–26.

33 Reid was an adventurer who easily made enemies, including Charles Barry, architect of the new House of Parliament, to which project Reid was assigned responsibility for the ventilating system. Reid spent the last years of his career in the United States. There is currently no satisfactory biography, though papers have been published suggesting that he was the designer of the world's earliest air-conditioning system (in St George's Hall, Liverpool).

34 Anderson, *Playfair Collection*, p. 11.

35 Brock, William H., *Justus von Liebig: The Chemical Gatekeeper* (Cambridge: Cambridge University Press, 1997), pp. 342–51 (Appendix 2: The British and American Network of Liebig's Pupils, Disciples, and Giessen Graduates); 116 names are listed.

36 Grant, Robert E., *On the Study of Medicine: Being an Introductory Address*

Delivered on the Opening of the Medical School of the University of London (London, 1833).

37   For statistics of total class size, see Morrell, 'Science and Scottish University Reform'; for Edinburgh medical graduations, see [University of Edinburgh, Faculty of Medicine] *Nomina eorum qui gradum medicinae doctoris in Academia Jacobi Sexti Scotorum Regis, quæ Edinburgi est, adepti sunt : ab anno MDCCV. ad annum MDCCCXLV...* (Edinburgh 1846).

38   Kay, John, *A Series of Original Portraits and Caricature Etchings*, 2 vols (Edinburgh, 1842), no. XCII, facing p. 448.

39   Robert G.W. Anderson [unpublished lecture notes], Society for the History of Alchemy and Chemistry lecture, Birkbeck College, 13 December 2008, 'From Union to Jubilee: the Southerly Drift'; Davies, Alwyn and Peter Garrett, *UCL Chemistry Department 1828–1974* (St Albans: Science Reviews, 2013).

ELEVEN

A Golden Cage, but Will the Birds Sing?':
William Gregory, Lyon Playfair
and Alexander Crum Brown

ANDREW J. ALEXANDER

Introduction

By the end of the eighteenth century, the University of Edinburgh had built a significant reputation for chemistry. This chapter discusses how chemistry flourished at the University during the nineteenth century. It would be impossible to cover a century in so few pages, so the discussion is limited to key characters from the period, reinterpreting what is known already and reporting new findings. For in-depth expositions of what has been uncovered of this period in the School's history, the reader is directed towards the works of Anderson[1] and of Doyle.[2] At his zenith, Thomas Charles Hope, who taught from 1795, had commanded the attention of audiences of over 500, but by his final session (1842–3) attendance had declined to 188.[3] Just before the 1843–4 session, Hope suddenly resigned. His written notes reveal the pressure of that last year:

> Progress 1842–3 I was <u>very</u> <u>much</u> rushed in the latter part of this course & had to abridge valuable matter [. . .] Ergo in 1843 push vigorously from the start – Specially abridge on Heat, which occupies too much of the Course.[4]

As is discussed below, with regard to Hope's successor, it is interesting to note that the subject of heat was singled out for abbreviation. Not only that, Thomas Stewart Traill, professor of medical jurisprudence, who became caretaker of the chemistry class for the session 1843–4, remarked that Hope's material on organic chemistry required expansion and updating.[5] This would be remedied by the three professors who followed Hope. Two students who were inspired in very different ways by chemistry teaching, Archibald Scott Couper (1831–1892) and Arthur Conan Doyle (1859–1930) are also treated in this chapter.

William Gregory

William Gregory (1803–1858) was appointed to the chair of chemistry in October 1844. The town council appointed him simply 'professor of chemistry' and did not include medicine in the title. He had been a pupil of Hope (1820–3), graduating as MD in 1828 with the thesis *De Principiis Vegetabilium Alkalinis* (concerning the principles of vegetable alkalis). In the years before his appointment he was an assistant to Jean Pierre Robiquet (Paris, 1827–8), Edward Turner (London, 1828–9), and Justus von Liebig (Giessen, 1835–6). He had also gained significant experience in teaching as a private lecturer at Edinburgh (1829–35) and Dublin (1836); he was appointed as Professor of Chemistry at Anderson's University in Glasgow (1837–8) and Professor of Medicine at King's College Aberdeen (1939–43).

Gregory rapidly became expert in organic chemistry, and by 1831 was delivering specialised courses in this area.[6] He became personally close to Liebig, and in 1839 published the first of several translations of Liebig's works, thereby bringing the leading chemistry of the era to a British audience. In 1845 he published his own textbook, *Outlines of Chemistry for the Use of Students*, which he prefaced:

> Every teacher of Chemistry must have felt the want of a compact text-book, the price of which might place it within the reach of every student; and it is the long-felt sense of this want which has led me to compile these outlines.[7]

Such sentiments still hold true today. He goes on to give 'some apology, or explanation at least' to his decision to omit certain subjects, in particular 'Heat, Light, Electricity, and Magnetism'. His reasoning was that not only should these subjects belong to physics, but that chemistry – particularly organic chemistry – had expanded in scale and importance. His *Handbook of Organic Chemistry* (1852) was the first specialised organic chemistry text by a British author, and he published a similar book on inorganic chemistry the following year.

In terms of research, Gregory is most noted for his methods of purification of morphine (1831) and chloroform (1850). Morphine had been identified by Friedrich Sertürner (1783–1841) in 1804, and commercial production of the acetate salt began by the 1820s; however, the quality was poor and the process expensive, requiring good grain alcohol. Gregory's method was to extract the morphine base from raw opium using ammonia, and to use hydrochloric acid to crystallise out the morphine hydrochloride salt.[8] The resulting process was quickly taken up by local pharmacists John Fletcher Macfarlan (1790–1861) and David Rennie Brown (1808–1875), and commercial production began in 1833. Gregory did not patent his process: Colman Green argues that at the time, there would have been no cognition of the value of pure morphine over laudanum, and the physiological effects of both were well known.[9] It was not until 1853, when the Scottish physician Alexander Wood (1817–1884) demonstrated the first use of the hypodermic syringe for pain relief, that purity of drugs would prove critical.[10] Several sources erroneously suggest that his wife Rebecca Wood (née Massey) became a morphine addict and died of an overdose. In fact she died on 6 February 1894 from acute pneumonia.[11]

Chloroform was introduced into medical practice as an anaesthetic in 1847 by James Young Simpson (1811–1870). Local companies, such as Macfarlan & Co., had put it into production by the following year. Production was also started in London, but it was often of questionable quality, with associated side-effects. Gregory developed a simple method that involved purification with sulphuric acid and manganese dioxide. In his paper to the Royal Society of Edinburgh (read 18 March 1850), he states:

> There are still, however, many makers in other places whose chloro-form is not so pure; and I shall now describe the method which, with Mr Kemp, I have employed for purifying [. . .] any commercial chloroform [. . .] a process which will enable any medical man to purify it for himself with the greatest facility.[12]

The simplicity of the method was a master stroke, at once enabling any stock of chloroform to be rendered safe and useful. Gregory states that the specific gravity of the best commercial chloroform at the time was 1.480, with samples specially prepared by chemists as high as 1.497. From a commercial sample, Gregory's process produced a liquid with gravity of 1.500.

Gregory was a strong believer in various pseudosciences, such as spiritualism and hypnotism. He wrote numerous letters on the subject, and even a short book.[13] There is a spiritual device, labelled as a 'tetragrammaton', in the Playfair Collection at the National Museums of Scotland, which Anderson explains must have belonged to Gregory.[14] In a letter to phrenologist George Combe, Gregory notes that phrenology played a part in deciding on his marriage to Lisett Barbara Scott; she was also wrapped up in the supernatural.[15] They had only one son, who was named James Liebig Gregory.

Gregory was noted to be of large build, and he was not in the greatest of health for much of his life, having suffered a fever in 1826 which was said to have been caused by his experiments. For this reason, in later years, Gregory turned his attention to less strenuous activities such as microscopy, from which he published several observations on the eukaryotic algae commonly known as diatoms.[16] Gregory's training in the best laboratories of Europe could have enabled him to build a school that rivalled the best in the world. Missed opportunities and fragile health, however, mean that he did not expand the teaching of chemistry in Edinburgh, particularly on the practical side. Nevertheless, he earned a healthy respect from his students, and was able to lecture on organic chemistry without notes. His character perhaps is best summed up in the following quote:

> The late professor was remarkable for his coolness and self-possession under circumstances which would have been, to say the least of it, trying to many. On one occasion, during his lecture, a tube of peroxide of chlorine burst in his hand, and a fragment of glass entering his eye, caused the aqueous humour to escape – nevertheless, he proceeded with his lecture and finished it.[17]

Lyon Playfair

Lyon Playfair (1818–1898) was appointed to the chair of chemistry in 1858, following the death of Gregory. He was educated at various Universities including St Andrew's (1833), Anderson's University (up to 1828, named Anderson's Institution) in Glasgow (1835), Edinburgh (1837) and University College, London (1838). He obtained his PhD at Giessen under Liebig in 1841. He had stood unsuccessfully as candidate for the chair at Edinburgh in 1844, at the tender age of only 26;[18] by 1858, despite having already embarked on a career in public service, the Edinburgh chair proved too tempting, and he stood again.[19] Because he was such a public figure, largely through his involvement with the 1851 Great Exhibition and his friendship with Prince Albert, a significant amount is known about Lyon Playfair, from the numerous records and correspondence that were handed down. Much of this can be found in the biography by Wemyss Reid.[20]

Playfair was a practical scientist, and he did much to develop and promote industrial and agricultural chemistry. His most significant contribution to pure chemistry was the discovery of nitroprusside salts (the sodium salt is still in use as a vasodilator in medicine). Playfair had a natural talent for understanding people, and knew how to obtain the best from students. He believed strongly in mentoring, and recognised the need for more postgraduate research. As special commissioner of the Great Exhibition of 1851, he created a scholarship scheme to attract talented students into research that is still running today; a number of these Exhibition fellows have gone on to become Nobel laureates. Playfair had an immense and lasting influence on the teaching of chemistry at Edinburgh. He instituted modern laboratory space, spending all of his income as professor from the first year (and a large fraction during subsequent years) on the equipment and running of the laboratories. He introduced tutorials, class merit certificates, and the marking of examinations by percentages. Playfair considered that the classes were too large for effective teaching; the tutorials were designed to drill students in exercises on the methods laid out in lectures. The merit certificates and medals were designed to encourage sustained effort and competition from the wider class.[21]

Up to the middle of the nineteenth century, professors of chemistry at Edinburgh privately hired any assistants that they needed. They certainly were needed: the annual course of chemistry lasted a gruelling six months, lectures being given on five days every week. Most lectures incorporated several demonstrations, and assistants had to organise for apparatus and chemicals to be on hand for the professor to perform these. Increasingly, as the new chemistry facilities became available from the 1820s, some students themselves undertook practical work, and this caused further strains on the system. The Universities (Scotland) Act of 1858 recommended that three assistants, employed by the University, be assigned to the Edinburgh professor. These were to have distinct responsibilities: one for preparing demonstrations, one for running the practical class, and the third for running the advanced laboratory. This change occurred at the very moment that Playfair was appointed to the Edinburgh chair. Several of the assistants hired in this way would end up having distinguished careers of their own, in particular James Dewar (1842–1923), who pioneered the liquefaction of gases.

The chemist appointed to be Playfair's second assistant, Archibald Scott Couper (1831–1892), had a brilliant but brief career, but thereafter a deeply tragic life. The reader is directed elsewhere for a full account;[22] only a brief summary of what is known will be given first, followed by some new findings. Couper was born in Kirkintilloch, educated in Glasgow and Edinburgh, and studied chemistry in Berlin. In August 1856 he moved to Paris to work with Charles Adolphe Wurtz (1817–1884). In early 1858, Couper handed a paper entitled *Sur une nouvelle théorie chimique* (On a new chemical theory) to Wurtz, for submission to the French Academy. Among other important features, the paper contained a statement that carbon was tetravalent. Wurtz was not a member of the Academy and passed it to another, thought to be Antoine Jérôme Balard (1802–1876), to gain leave to present it. For some reason a delay occurred, and Couper's paper was not read until 14 June 1858 by Jean-Baptiste Dumas (1800–1884). Meanwhile, a paper by August Kekulé (1824–1896), dated 16 March 1858, was printed in Liebig's *Annalen* on 19 May, also containing a statement of the tetravalence of carbon.

Thus, Couper had lost priority for establishing the tetravalance of carbon, and Kekulé was quick to criticise him severely for it.[23] An argument between Couper and Wurtz resulted in Couper being told to leave the laboratory. Couper's movements after June 1858 are unclear, until a letter from Lyon Playfair to Couper, dated 4 December 1858, reveals:[24]

> I have written to Professor Bunsen strongly expressing my desire to engage you and asking him to put a stop to any arrangements he may have been making for me.

Playfair then offered Couper the position of second laboratory assistant, to commence from 2 January 1859. From the above statement it seems that Playfair had supported Couper in obtaining a place in Robert Bunsen's laboratory at Heidelberg (Bunsen and Playfair were close friends). Now, in need of an assistant, he was keen to hire Couper himself. We know that Couper took the position from an account by Greville Williams, one of Playfair's other assistants. However, by May 1859, Couper had had a mental breakdown and had been admitted to a private mental institution, Garngad House, in Glasgow. He was a patient at similar institutions until November 1862, when he was released into the care of his mother. He never returned to chemistry. He died 11 March 1892, his mother surviving him.

The events as summarised above may be found in the article *The Couper Quest* (1934) by Leonard Dobbin.[25] However, the following questions remain unanswered: (i) Do documents and effects belonging to Couper still exist? (ii) Why did Playfair go out of his way to help Couper? (iii) What caused the mental breakdown? In his article, Dobbin mentioned a box of documents labelled 'Archibald Scott Couper' handed to him by the family of Alexander Crum Brown. The documents could not be located in the University of Edinburgh School of Chemistry archives, but were eventually traced to the Chemical Heritage Foundation, Philadelphia.[26] The collection consists mostly of original correspondence between Richard Anschütz, Alexander Crum Brown and others, who traced Couper's history; excerpts from the correspondence are printed in Dobbin's article. In 2013, a living descendent of Couper – Mr John Watt Dollar – was traced, who searched the surviving family archives. The only surviving documents were an original reprint copy of Richard Anschütz's biographical work on Couper, and a photographic print (c.1890) of Couper's mother, Helen Couper (née Dollar).

How Playfair became acquainted with Couper is not clear. The letters of Playfair that survive are mostly those of political or scientific significance. At Heidelberg University there appears to be no correspondence between Bunsen and Playfair that might shed light on the matter. It should be noted, however, that Playfair had a deep sense of care for students, and would likely have recognised the value of Couper's work.

Regarding the cause and extent of Couper's breakdown, it may be tempting to connect his decline to the incidents in Paris. Albert Ladenburg (1842–1911), who worked in Wurtz's laboratory, stated that 'Couper seems to have [taken the affair] very much to heart, and in Paris it was therefore believed to date the beginning of his illness.'[27] Couper was admitted to Garngad House as a private patient on 15 May 1859, discharged 14 July, and then on 27 July that same year admitted to Saughton Hall Asylum, Edinburgh. He was transferred on 21 September 1860 to the Glasgow Royal Asylum for Lunatics at Gartnavel. His case notes from the Gartnavel hospital still survive: an excerpt from 10 October 1860 reads:

> Answers questions civilly but in a somewhat abrupt style [...] Tears books and papers, without any conceivable motive. When interrogated as to his conduct, replies that he is compelled to do it by some irresistible mandate.[28]

Couper's violent attitude to written papers might have resulted from anger towards science and learning. According to consultant psychiatrist Dr Allan Beveridge, who recently reviewed the historical case notes, Couper's condition was paranoid schizophrenia. Although it is possible that the events of Paris exacerbated Couper's mental decline, it is more probable that this condition was more deeply rooted, and that there may have been signs of it in his past. In a letter of 1906, Gustav Berring, a friend of Couper, wrote of their time together studying in Germany: '[Couper] was not, however, in robust health and always had to be concerned about guarding it.' Whether this was an allusion to Couper's mental state is not clear. One practical issue that does not appear anywhere in the medical case notes is mention of sunstroke as being a cause of Couper's illness, as has been discussed elsewhere.[29] The story about sunstroke may well have been circulated by the family as an explanation for the deeper underlying illness.

This sad story is concluded with a quote from Archibald Wilson of Kirkintilloch, who recalled overhearing a conversation between the young Couper and his mother:

> One morning Mrs Couper said to her son that she had had an extraordinary dream about him during the night. She dreamt that he was going to become one of the most famous men of his time. It must have been at a time when pessimism had got hold of him, for after a few minutes thought he replied 'Mother, my pirn is a' wound up, and I will never be anything more in the future than I am to-day'.[30]

Alexander Crum Brown

Alexander Crum Brown (1838–1922) succeeded to the chair in 1869 (Playfair resigned to return to public life in London), having previously been an extramural lecturer from 1863. He was born and educated in Edinburgh, obtaining an MA in 1858 and MD in 1861. He was the first candidate to take the DSc of the University of London in 1862. Like his two predecessors in the chair, Crum Brown studied in Germany: under Robert Bunsen (1811–1899) at Heidelberg, and Hermann Kolbe (1818–1884) at Marburg.[31] Crum Brown was a polymath and a polyglot; he was reputed to be, in his day, the greatest Chinese scholar in Great Britain.[32]

During his tenure, Crum Brown made several significant contributions to chemistry, although he is arguably not as broadly recognised as he ought to be; this muted recognition stems in part from his quiet, self-effacing character. Through loyalty to the society, he tended to publish in the Royal Society of Edinburgh journals, which were not as widely read as other European titles. Crum Brown is perhaps most famous for his contribution to structural chemistry. He devised the ball-and-stick representation for drawing molecules, and used it to explain that ethene consisted of two carbon atoms joined by a double bond. In 1864 he established the concept of structural isomerism: molecules with the same number and type of atoms can be arranged to form different structures.[33]

Crum Brown's talents included a passion for knitting. As a child, he invented a knitting machine, and later experimented with new methods of knitting three-dimensional interlocking surfaces. After his retirement, during the First World War, he was often to be seen riding on the trams in Edinburgh, knitting socks for the soldiers. Many years before the determination of the structure of sodium chloride by William Bragg and his son Lawrence, by X-ray diffraction techniques, Crum Brown built a model using knitting needles and balls of wool; the Braggs later received the Nobel Prize in 1915 for their work. Crum Brown's 1883 model still exists today in the Museum of the School of Chemistry.[34]

On 8 December 1868, a group of Edinburgh medical men met at the house of Dr Robert Blair Cunynghame where, 'after considerable conversation', the Round Table Club was formed (Plate 14). One of the 12 original members was Alexander Crum Brown, and another was Joseph Bell (1837–1911), who was subsequently appointed secretary. The main objective of the club was 'an occasional meeting simply for the purpose of good fellowship'. The Round Table Club constitution was drawn up, allowing for 20 members, and a set of seven by-laws was adopted, including:

5. Drink. Every member who orders drink shall pay for the same before leaving the hotel. 6. Matrimony. Every member on being married shall stand two bottles of whisky, or pay an equal value in money. 7. Professorship. Every member who becomes a professor shall stand champagne ad libitum, or pay two pounds in money.

The antics of the subsequent meetings, spanning the period to 1895, are recorded in *The Annals of the Round Table Club*.[35] Of the members of the Round Table Club, Joseph Bell stands out as the inspiration for the famous fictional detective, Sherlock Holmes, in the stories written by Arthur Conan Doyle. In an interview in 1892, Doyle noted that 'Sherlock Holmes is the literary embodiment, if I may so express it, of my memory of a professor of medicine at Edinburgh University.'[36] The uncanny ability of Dr Bell to deduce – seemingly from thin air – information about his patients, became the basis for Holmes's deductive powers as a detective.[37]

The first Sherlock Holmes story (published in 1887) was *A Study in Scarlet*, in which Dr John Watson is introduced to Holmes as a potential room-mate by Stamford, an acquaintance of both. The introduction takes place at the chemical laboratory where Holmes is working. At this meeting, Holmes says 'I generally have chemicals about, and occasionally do experiments. Would that annoy you?' Later on, Watson enumerates Holmes's strengths and weaknesses, among which he lists:

5. Botany. – Variable. Well up in belladonna, opium, and poisons generally. Knows nothing of practical gardening.
6. Knowledge of Geology. – Practical, but limited [. . .]
7. Chemistry. – Profound.
8. Anatomy. – Accurate, but unsystematic.[38]

It is clear that Doyle created Holmes with chemistry as one of his key talents. Where, then, did Sherlock Holmes learn chemistry? Or rather, where did Doyle learn chemistry? As is well documented, Doyle studied medicine at Edinburgh University from 1876 to 1881. Medical students at that time were required to take courses in chemistry and practical chemistry, and Doyle took both during his first year. These included 100 lectures on chemistry with Professor Alexander Crum Brown, and practical chemistry under Dr Andrew Peebles Aitken (1843–1904), who was laboratory demonstrator and the professor's assistant. Doyle's entry in the Chemistry Class Register records matriculation ticket number 288, at 2 Argyle Park Terrace, where he was living with his parents and siblings.[39] At this time, the chemistry department was still in the south-west corner of the Old College, as designed by architect William Henry Playfair. With regard to the practical chemistry class, according to George Stephenson (writing under the pseudonym of 'Alisma'):

> The class was rather uninteresting, and the teaching consisted mainly in the detection of various metals and poisonous substances in fluids, such as mercury, antimony, lead, arsenic, etc., etc.[40]

Doyle's examination results are revealing. In the written part of his first professional examination, taken in April 1878, Doyle received grades of 'S' (Latin *satis* – sufficient) for Chemistry, 'S+' (above sufficient) for Botany and Natural History, and 'B' (*bene* – good) for Chemical Testing.[41] On his subsequent oral examination he received 'S' for Chemistry, having been tested on 'Acids of nitrogen. Nitric acid. N_2O_5. Nitric acid + metals'. So it is clear that Doyle had a good grasp of chemistry, particularly the practical side. The Sherlock Holmes stories, collectively referred to as the Canon, consist of four novels and 56 short stories. Not including other sciences, chemistry appears in some form in 57% of the Canon.[42] Topics include carbon monoxide and dioxide; coal-tar derivatives; inorganic acids; acetones; phosphorus; barium bisulphate; amalgams; and various drugs and poisons.[43] We can compare the chemistry that appears in the Canon against surviving lecture notes of 1890, written by John Alan Murray, and a personal notebook of lecture demonstrations written by Crum Brown from 1861 to 1870.[44] Approximately 60% of the chemistry contained in the Canon was covered in the chemistry lectures.

Coal gas (often called town gas or illuminating gas) contains hydrogen, carbon monoxide and a mix of hydrocarbons in varying proportions. Gas lamps were used widely in Britain for illumination in the nineteenth century. During Doyle's lifetime, the easy access to poisonous chemicals such as carbon monoxide resulted in commonplace suicides, murders and accidents. In the Canon, there are deaths by carbon monoxide poisoning described in *The Greek Interpreter* (published in 1893) and *The Adventure of the Retired Colourman* (published in 1926). The reactions of the 'carbonic acid gases', formation and physiological effects were covered at various points in twenty of the lectures. Crum Brown described carbon monoxide as 'medically poisonous'. In *The Greek Interpreter* it is the incomplete combustion of charcoal that poisons the Greek man being held captive, while the interpreter barely escapes with his life. In Lecture 11 of Murray's notes, we find that 'carbonic oxide is carbon half-burnt' (Plate 15):

> Formation of Carbonic Oxide in a Fire. In a glowing fire air enters at [the] bottom, mixes with hot carbon and forms carbonic acid gas [CO_2]. As this passes up it is changed into carbonic oxide [CO]. This burns with a blue flame above the coals. This formation of carbonic oxide is very important medically. Poisoning due to fumes of charcoal is due to the carbonic oxide in these fumes.[45]

The 40% of chemistry mentioned in the Canon that was not covered in the chemistry lectures can be classed under the heading of drugs and poisons. In 1868, Alexander Crum Brown and Thomas Richard Fraser, another original member of the Round Table Club (Plate 14), published their first paper relating the physiological action of drugs to the chemical constitution, or structure, of the drug molecules. Their paper deals with salts of the alkaloids: strychnine, brucine, thebaine, codeine, morphine and nicotine. This was ground-breaking research, which showed that even minor structural changes could drastically change the action of the drug. For example, they found that strychnine methiodide is nearly 300 times less active than strychnine hydrochloride in causing death in rabbits. The work culminated in the award of the Royal Society of Edinburgh's Makdougall–Brisbane Prize to Crum Brown and Fraser.

In the 1878–9 session, Doyle took Professor Fraser's course in materia medica (pharmacology) consisting of 100 lectures. In his second professional examination, Doyle received above-average 'B' (*bene*) grades in both the written and oral parts of the exam. Doyle was surely aware of Crum Brown and Fraser's pioneering work. In *The Sign of Four* (published in 1890) Holmes and Watson study the dead body of Bartholomew Sholto, who has been shot with a dart. Holmes asks Watson as to his conclusion on finding the extreme rigor mortis; Watson replies, 'Death from some powerful vegetable alkaloid [...] some strychnine-like substance which would produce tetanus.'

Many of Crum Brown's lecture demonstrations were loud or malodorous. In the fifth lecture of his notebook we find a demonstration of ignition:

A large glass vessel was filled with CO_2 and a small quantity of gun cotton was placed within its bottom. A taper (lighted) was then dipped into the CO_2 with the view, if practicable, of exploding the cotton. The taper was instantly extinguished by the CO_2. A stout iron rod heated to a red heat was then inserted into the CO_2 touching the gun cotton when the cotton was instantly exploded.[46]

This description can be somewhat enriched by Doyle's fond memory of his former professor, Crum Brown, as written in his 1924 autobiography:

There was kindly Crum Brown, the chemist, who sheltered himself carefully before exploding some mixture, which usually failed to ignite, so that the loud 'Boom!' uttered by the class was the only resulting sound. Brown would emerge from his retreat with a 'Really, gentlemen!' of remonstrance, and go on without allusion to the abortive experiment.[47]

In Doyle's chemistry class there were 276 students, and by 1882–3 this had risen to 382. The expansion of the medical school, and growth in science degrees in general, led the University to build the new medical school buildings at Teviot Place. The Chemistry Department moved during the winter session, 1884–5. Upon opening the new laboratories, Crum Brown is reported to have said of them: 'A Golden Cage, but will the birds sing?'[48] In the latter part of his career, Crum Brown was affectionately nicknamed 'Crummie' by his students. Throughout his career, he taught and inspired many students who would go on to become noted scientists and physicians, such as Prafulla Chandra Ray, known as 'the Father of Indian Chemistry'. One of the students in the advanced laboratory class of 1884 was James Walker, who was to succeed Crum Brown in the chair of chemistry from 1908. The history of the twentieth century is best left, however, to the quatercentenary in 2113.

Acknowledgements

The author thanks the following people: John Dollar, descendent of Archibald Scott Couper; Dr Ivan Ruddock; Dr Allan Beveridge; Staff at the Centre for Research Collections at Edinburgh University Library; and Dr Robert Anderson. I am especially grateful to Professor Robert Donovan for his support during the tercentenary year, and for checking the Bunsen archives; and to Elizabeth Clarence, for her work on the chemistry of Sherlock Holmes, and for many useful discussions. We gratefully acknowledge an award from the Moray Endowment Fund of the University of Edinburgh.

Notes and References

1 Anderson, R.G.W., *The Playfair Collection and the Teaching of Chemistry at the University of Edinburgh 1713–1858* (Edinburgh: Royal Scottish Museum, 1978).

2 Doyle, W.P., *Scottish Men of Science* (Edinburgh: University of Edinburgh History of Medicine and Science Unit, 1982).

3 Anderson, *The Playfair Collection*, p. 39.

4 Edinburgh University Library Centre for Research Collections, Notes of Thomas Charles Hope, GB 237 Coll-12 Gen. 270.

5 Anderson, *The Playfair Collection*, p. 42.

6 Ibid.

7 Gregory, W., *Outlines of Chemistry for the Use of Students* (London: Taylor and Walton, 1845).

8 Gregory, W., 'A Process for Preparing Economically the Muriate of Morphia', *Edinburgh Medical and Surgical Journal* 35 (1931), pp. 331–8.

9 Green, G.C., 'William Gregory, M.D., F.C.S.: 1803–1858', *Nature* 157 (1946), pp. 465–9.

10 Wood, A., 'New Method of Treating Neuralgia by the direct application of Opiates to the painful Points', *Edinburgh Medical and Surgical Journal* 82 (1855), pp. 265–81.

11 Wood, Rebecca, GROS 685/05 0165: Statutory Register of Deaths, Scotland.

12 Gregory, W., 'Notes on the Purification and Properties of Chloroform', *Proceedings of the Royal Society of Edinburgh* 2 (1850), pp. 316–24.

13 Doyle, *Scottish Men of Science*.

14 Anderson, *The Playfair Collection*, pp. 123–4.

15 Doyle, *Scottish Men of Science*.

16 Miller, W.A., 'Report from the President and Council', *Quarterly Journal of the Chemical Society of London* 12 (1860), pp. 166–76.

17 Alison, W.P. , 'Account of the Life and Labours of Dr William Gregory', *Proceedings of the Royal Society of Edinburgh* 4 (1862), pp. 121–2.

18 Doyle, *Scottish Men of Science*.

19 Doyle, *Scottish Men of Science*.

20 Reid, W., *Memoirs and Correspondence of Lyon Playfair* (London: Cassell & Company, 1900).

21 Doyle, *Scottish Men of Science*.

22 Dobbin, L., 'The Couper Quest', *Journal of Chemical Education* 11 (1934), pp. 331–8; Anschütz, R., 'Life and Chemical Work of Archibald Scott Couper', *Proceedings of the Royal Society of Edinburgh* 29 (1909), pp. 193–273; Duff, D.G., 'A.S. Couper: the forgotten genius', *Chemistry in Britain* 23 (1987), pp. 350–4.

23 Kekulé, A., 'Remarques de M. A. Kekulé à l'occasion d'une Note de M. Couper sur une nouvelle théorie chimique', *Comptes Rendus* 47 (1858), pp. 378–80.

24 Dobbin, 'The Couper Quest'.

25 Miller, 'Report from the President and Council'.

26 Chemical Heritage Foundation, Philadelphia, PA: Archibald Scott Couper original materials, 2010.013/1/11–16.

27 Anschütz, 'Life and Chemical Work of Archibald Scott Couper'.

28 NHS Greater Glasgow and Clyde Archive, Mitchell Library, Glasgow: Archibald Scott Couper, HB13/6/6 (patient register), HB13/5/65 (case notes),13/7/67B (admission warrants).

29 Duff, 'A.S. Couper'.

30 *Kirkintilloch Herald*, 'The Couper Centenary', 14 October 1931, p. 5.

31 Doyle, *Scottish Men of Science*.

32 Stephenson, G.S., *Remiscences of a Student's Life in At Edinburgh in the Seventies (by Alisma)* (Edinburgh: Oliver and Boyd, 1918).

33 Doyle, *Scottish Men of Science*.

34 University of Edinburgh: School of Chemistry Archives.

35 McKendrick, J.G. (ed.), *Annals of the Round Table Club* (Stonehaven: John Taylor & Co., 1908).

36 Blathwayt, R., in *The Bookman* (London: Hodder & Stoughton, 1892), pp. 50–1.

37 Doyle, A.C., *Memories and Adventures* (London: Hodder & Stoughton, 1924).

38 Doyle, A.C., *A Study in Scarlet* (London: Ward, Lock & Co., 1887).

39 Edinburgh University Library, Centre for Research Collections: Lectures on Chemistry (1873–85), GB 237 EUA IN1/ACU/C2/9.

40 Stephenson, *Remiscences*.

41 Edinburgh University Library, Centre for Research Collections: Graduates in Medicine, GB 237 EUA IN1/ADS/STA/8/1881.

42 Clarence, E.A., M.Chem. Thesis, University of Edinburgh, 2013.

43 O'Brien, J., *The Scientific Sherlock Holmes* (Oxford: Oxford University Press, 2013).

44 University of Edinburgh: School of Chemistry Archives.

45 Ibid.

46 Ibid.

47 Doyle, *Memories and Adventures*.

48 Birse, R.M., *Science at the University of Edinburgh 1583–1993* (Edinburgh: University of Edinburgh, 1994).

Afterword

HASOK CHANG

The name of Joseph Black is what stands out to most casual observers of the earlier periods of history of chemistry in Edinburgh, or even in all of Scotland. As the papers collected in this volume demonstrate abundantly, there has of course been much more than Black's work in the history of Edinburgh chemistry – 'that celebrated school', as Thomas Thomson put it in the mid nineteenth century.[1] This collection presents a very rich tapestry of historical learning and insight about that celebrated school of chemistry. Here I would like to take a slight step back, and try to discern the main characteristics of Edinburgh chemistry, and the main factors that allowed it to flourish as it did, especially during the Enlightenment period. What was required for the doing of chemistry in that period? Were conditions in Edinburgh particularly favourable for it? I hope that the sketchy picture that I am going to paint will be helpful in setting the more learned and detailed accounts in the rest of the book into a kind of context that will make them more meaningful, especially to non-specialist readers.

Joseph Black

It is useful to begin with a simple and unsophisticated picture of what was so great about Edinburgh chemistry. Here, even through a pair of eyes educated by the erudition contained in this volume, comes Black again.[2] What contemporaries marvelled at and posterity remembered was his pioneering work in two areas: in pneumatic chemistry, he demonstrated that combination with fixed air (in modern terms, carbon dioxide, CO_2) changed the properties of substances in characteristic ways; in the study of heat, his work made crucial steps towards the new concepts of latent and specific heat. We need to remember what momentous discoveries these were – this is something that expert historians often lose sight of, in the midst of their erudition. The recognition of scientific significance is not incompatible with historiographical sophistication; reading Black's work in the context of larger currents in the development of chemistry should heighten, not reduce, our scientific admiration of it.

Black's work on fixed air formed a paradigm of work in pneumatic chemistry.[3] Certainly the gases emerging from various chemical processes had been studied by other chemists before, including Jan Baptist van Helmont (1580–1644), John Mayow (1641–1679) and Stephen Hales (1677–1761). Helmont had in fact studied fixed air more than a century before Black, calling it 'gas sylvestre' (he also coined the word 'gas' itself). What was truly innovative about Black's work was his clear recognition that gases could be put *into* (in other words, 'fixed' in) other chemical substances.[4] This work brought gases (or, 'airs', as they were more commonly called then) firmly into the realm of the budding tradition of 'compositionist chemistry' that would soon find clear expressions in the work of Antoine-Laurent Lavoisier (1743–1794) and then John Dalton (1766–1844). In the compositionist tradition, basic chemical substances were seen as unalterable building blocks, whose combination and re-combination constitute chemical reactions.[5] However, Black's work also had a great affinity to the older tradition of 'principlist' chemistry (in which the theory of phlogiston is often placed), because in his view, fixed air behaved like a 'principle': its addition to various substances changed their properties in characteristic ways (in the cases he was investigating, removing alkaline causticity). Therefore Black's work occupied a very subtle position at the crossroads of the Chemical Revolution, which he helped to bring about in important ways, and slowly and reluctantly joined. (John Christie's contribution to the present volume discusses further the place of Black and others in Edinburgh, including students, in the Chemical Revolution.)

Black's work on heat, though originating from a different direction of work, had the same kind of intellectual valency. The basic outlines of this work are well known, though it is very difficult to be certain about Black's precise reasoning, as his work on heat was never published by himself; his ideas spread mostly through his lectures and through the works of his pupils, some of whom, most notably William Irvine, developed their ideas in very different directions from Black's own.[6] It seems reasonably certain that Black's work on specific and latent heat was conceived in a chemical manner, like Lavoisier's later, seeing heat as a substance (of a rather unspecified nature) capable of chemical combination with other substances. It was not Black's style to proceed on the basis of a rigid theoretical plan, and it is interesting to see how different theoretical elements were carefully and undogmatically brought in to guide and frame his experimental research.

Regarding specific heat, Black furthered the notion that each substance had a characteristic affinity for heat, and made precise measurements of that affinity in terms of how much heat a unit amount of the substance was able to absorb as it went up one degree in temperature. This is not exactly like other chemical operations, but there is sufficient similarity here with standard chemical investigations of the day, including Richard Kirwan's affinity measurements and J.B. Richter's stoichiometric investigations.[7] Black's work on latent heat was even more closely entwined with chemical reasoning. The structure of his reasoning concerning latent heat is very much parallel to the structure of his reasoning concerning fixed air, and this pattern of thinking was eagerly adopted by Lavoisier, who reified heat more readily (giving it the name of 'caloric'), and treated latent heat as caloric chemically combined with another substance. Here, in Lavoisier's thinking[8] as well as Black's, there was an uneasy mix of compositionist and principlist intuition: latent heat could almost be treated as a chemical building block of substances (and its quantities subject to precise measurement, both for Black and Lavoisier), but its main chemical function was a principlist one: to impart fluidity to the substances with which it combined.

The Edinburgh Chemical Tradition

Important as Black's research achievements were, his legendary chemistry lectures at the University of Edinburgh were even more widely known. If there was a continuity of tradition in Edinburgh chemistry, it was not so much in the specific content of doctrine or in the particular directions of research, but in the methods and customs of pedagogy. The brilliance of chemistry teaching in Edinburgh was an arc continuing from William Cullen through Thomas Hope and William Gregory, as discussed by various contributions in the present volume.[9] Hundreds of students flocked to these lectures every year, and a small number of them went on to become established chemists (aside from those who became doctors, pharmacists, etc.). It is, however, not entirely clear why it would have been considered so important for medical students in that period to undertake these chemistry courses, which focused on aspects of basic chemistry that would have been quite irrelevant for daily medical practice.

The common picture of the Edinburgh chair of chemistry in the early years is that of a dynasty which produced one great king in the middle of it, whose glory gradually faded out after that – by 1833 the zoologist Robert E. Grant, himself an Edinburgh graduate, lamented the state of the University of Edinburgh: 'that University has sunk'.[10] This picture is significantly revised and enriched in the papers presented in this volume; they helpfully avoid attributing success to individual geniuses, instead seeking to identify the institutional, methodological, cultural and material resources that enabled the success.

John Powers gives a detailed and nuanced account of the influence of Herman Boerhaave (1668–1738) on the initial setup of medical and chemical teaching in Edinburgh; James Crawford, the first occupant of the Edinburgh chair from 1713, had himself studied briefly in Leiden, and was strongly influenced by Boerhaave's pedagogical framework. Georgette Taylor gives an instructive comparative view of the celebrated William Cullen and his little-known predecessor Andrew Plummer. These are valuable contributions enriching our view of the early days of Edinburgh chemistry. They lead us into the late eighteenth century, during which chemistry flourished along with general intellectual learning and culture in Edinburgh. Addressing this fertile period, John Henry clarifies the context of the Scottish Enlightenment and Newtonianism for Edinburgh chemistry. John Christie gives an insightful account of the place of students in chemical discussions, as well as the changing fortunes of phlogiston chemistry. Matthew Eddy explores Black's use of diagrams for pedagogical purposes. All this is followed by Robert Anderson's account of Black's successor, Thomas Charles Hope; even though Hope's lectures were highly popular and up to date at least for some time, Hope embodied the post-Black tradition in Edinburgh that emphasised teaching more than research. After Hope's long tenure ended in 1843, what was a distinct tradition of chemistry in Edinburgh began to develop in rather different directions, as shown in Andrew Alexander's survey of developments in the latter half of the nineteenth century.

If the question is what was *distinctive* about Scottish chemistry, especially Edinburgh chemistry, I think the answer is, indeed, teaching and what followed from teaching. It may be correct, after all, to identify Edinburgh chemistry as a dynasty of University professors as teachers, more than by thinking about the specific content of the research conducted by the professors, their associates and their students. It is instructive to draw a contrast here with London, or the English provinces, or Paris, or the various German universities. The university setting in Scotland, while close enough to what we might take for granted today, was nearly unique in chemistry up to the early nineteenth century among the leading sites of chemistry. In London there were no universities until the 1820s, and chemical research and communication there were carried out by those belonging to the medical schools located at hospitals, or by amateur researchers and private lecturers, and then most famously at the Royal Institution from 1800. It is interesting to note that when the concept of the university came to London in the form of the University of London (later re-named University College London as the dust settled from the founding of rival King's College), the first four professors of chemistry came from Scotland: Edward Turner, Thomas Graham, Alexander Williamson and William Ramsay.[11] In the English provinces the situation was similar, with the dissenting academies playing a significant role until the founding of universities in Manchester, Durham, etc. Even in Paris with its well-established and great university, chemistry was not a focus of instruction and research; that would only change significantly with the founding of the École Polytechnique and other institutions in the Revolutionary period.[12] The situation was also similar in the various German universities, though that would also begin to change rapidly during the period of the Chemical Revolution.[13] In Oxford and Cambridge chemistry was taught but not examined, and neither teaching nor research of particular distinction seems to have happened at either place until well into the nineteenth century.[14] The cultural contexts there were also very different from the one prevailing in Edinburgh, with its ferment of the Enlightenment. The only clear model that the University of Edinburgh drew from seems to have come from Leiden, as John Powers's contribution to this volume elaborates.

It is clear that the teaching of chemistry was taken very seriously in Scotland by the leaders of the field. This was not only evident at Edinburgh but also Glasgow, which only managed to retain Cullen and then Black for a short while, but later benefited from Thomas Thomson's long service from 1817 to 1841.[15] But exactly what kind of teaching was happening in Edinburgh chemistry? I think it was very important indeed that students' fees were paid directly to the professors (who might even be unsalaried) in the Scottish system during the glory days of Edinburgh chemistry. This provided a clear motivation to hold large lectures that were attractive to a diverse population of students. The legendary practical demonstrations by Black and other Edinburgh professors had a perfectly clear motivation, then, as did the effective visualisation techniques discussed in Matthew Eddy's paper. Similar things also did happen at the Royal Institution in London and at various similar institutions, but they were not university settings. Edinburgh therefore constituted an important pioneering site of outward-looking academic chemistry teaching. At the same time, it is understandable that there was little success in intensive practical training in this setup, the kind that would prove a powerhouse of chemical research in settings such as Justus von Liebig's school in Giessen.[16]

What about Research?

There isn't a great deal of discussion in the present volume about the chemical *research* that took place in Edinburgh. It is good that historians now pay proper attention to teaching, but we should not lose sight of research, either. Robert Anderson's paper makes clear that Hope, for one, saw a clear difference between research-focused chemists and teaching-focused chemists. But it is also clear that on the whole, significant chemical research and teaching both did take place in Edinburgh. Then we may ask: did the research work of the early Edinburgh chemists benefit from their teaching work? This seems to have worked out wonderfully for Black and perhaps a bit for Cullen, but not for the other professors. We know that even Black did not *publish* very much, and that was not unique to the Edinburgh setting. Henry Cavendish comes to mind as another eminent research chemist who published very sparingly.[17] At least in parts of Britain there was something very far from today's 'publish or perish' academic culture. Today's harried academic may justly wonder whether these greats would ever have got tenure in a modern American university, or how well they would have survived the research-assessment exercises in the United Kingdom. And was scientific research any the worse in this relaxed publishing culture?[18] Setting that question aside, it is easy enough to imagine that in the academic economy and culture of Enlightenment Edinburgh there would have been a strong incentive to focus on the delivery of teaching rather than research. Some, such as Black, were able and willing to do the additional work needed to maintain a productive research career, others not.

In assessing this situation there is an inherent difficulty for the historian. Looking at an academic culture that is not bent on publishing, how do we get a sense of what unpublished research went on? This is a difficulty that historians know very well how to handle by going to archival and circumstantial sources, but handling it does take a considerable amount of work and ingenuity. Here we may have much to learn from the historians of alchemy. We may also have much to learn from archaeologists, and even work directly with them. The instruments and materials that can be (sometimes literally) dug up will surely give us a very useful glimpse of non-teaching-related research, as indicated in Tom Addyman's paper on studying the material legacy of the chemical work found in the early chemistry stores recently excavated at the Old College Quadrangle of the University.[19] In a manner more familiar to historians, insights can be gained from the intentionally preserved material legacy, such as the collections of chemical apparatus in the National Museums of Scotland, described in Alison Morrison-Low's contribution. Similar enlightenment may come from studying buildings, as suggested in Peter Morris's detective work trying to locate Black's house.

In assessing the research contributions of Edinburgh chemists, there are some interesting comparisons and contrasts we could make. Thomson in Glasgow again comes to mind, as an early nineteenth-century Scottish professor who was more research-active than his counterparts in Edinburgh. Was this due to the years he had spent in London? Was Thomson's research greatly helped by his decision to train students practically in a teaching laboratory? Or we might make a comparison with the highly research-active Scottish professors of chemistry at UCL in the first half of the nineteenth century, namely Turner, Graham and Williamson. An important factor here may be that chemistry at the University of Edinburgh remained strongly attached to the teaching of medicine, in contrast to the situation at these other universities. Thomson was the Regius Professor of Chemistry, while Hope was the Professor of Medicine and Chemistry. The difference in titles may well have been a reflection of different institutional realities.

Why did Edinburgh Chemistry Flourish around the Time of Black?

I began by asking what enabled such flourishing of chemistry in Edinburgh in the Enlightenment period. So far I have based my discussion on the specific set-up of teaching in Edinburgh. Now it is time to consider the larger contexts, both philosophical and political, which are very interestingly correlated with each other.

The Scottish Enlightenment is clearly an important backdrop here,[20] and the ideology of the Enlightenment included Newtonianism, which is discussed in John Henry's paper. But it seems to me that Newtonianism was often a matter of lip-service, or rather a widely shared ideology convenient for a variety of purposes (more Protean than phlogiston). So many people in the Enlightenment period were 'Newtonian' by self-designation and in the rejection of unfounded authority, whether religious or classical. The anti-authoritarian tendency can also be seen in the active roles played by Edinburgh students and the independent views held and defended by them, as discussed by Georgette Taylor and John Christie. But all this is not unique to Newtonianism proper, nor is it special to Scotland; it is merely a very general characteristic of the Enlightenment.[21]

That general characteristic of anti-authoritarian empiricism, however, *is* important for our story. The real contributions of Edinburgh chemistry came through an operational sort of affinity theory, and later, an operational sort of pneumatic chemistry. This can be seen clearly in Black's work, and in William Cullen's work alluded to in Georgette Taylor's paper. The same theme is also brought out very interestingly in John Christie's discussion of the operational reality of phlogiston in Priestleian eudiometry, in the phlogistic economy of nature, etc. The important aspect of the basic outlook on science that I am pointing to is not positivism, as has sometimes been suggested. Rather, what I am trying to capture by the term 'operational' is the pragmatist sense of developing the sort of knowledge based on concrete engagement with the material processes of nature. This, I would argue, is a lasting element of the Edinburgh tradition that had significant impact far outside Edinburgh, too. So, the important legacy here is not so much that of Newton, but more that of Herman Boerhaave, who turned elements/principles into 'instruments' as discussed by John Powers. In this connection, it would be very informative to explore further the connection between industry, pharmacy and agriculture, which Alison Morrison-Low and Andrew Alexander touch on.[22]

Why Should Chemists Care?

I would like to close with some reflections on a different kind of question, at least giving recognition to the wonderfully mixed audience of historians, chemists and others that were assembled at the conference from

which this volume arises. Why should the working scientists pay attention to all this history? What do they gain from the study of history? Is it
only a matter of curiosity and amusement, and some vague sense of their
own heritage and context? All that, actually, is not to be dismissed. But
are there also lessons in this history, or any other history? While being
somewhat wary of this business of 'learning from history', I will venture
some very brief thoughts.

Taking in the various phases of work that have taken place in
Edinburgh chemistry over the centuries reminds us that different kinds
of teaching may bring different kinds of successful learning. For example,
it is interesting to note the usefulness of what John Christie calls 'aggressive or agonistic emulation'. More generally, it is instructive for scientists to learn that science can be, and has been, successfully organised in
many different ways. History can illustrate how various successful strate-
gies of research, training and communication can be devised in order
to adapt science to changing circumstances. Similarly, although we have
not entered into this issue deeply here, it is also very useful to learn that
various conceptual and practical schemes (even phlogiston chemistry)
can be, and have been, employed in successful scientific work. This
points to the mind-opening and pluralist function of history. If we can
let history teach us the openness of letting the past show us the variety of
the meanings and methods of success, the New Scottish Enlightenment
advocated by Sir John Arbuthnott may indeed draw useful inspiration
from the original one.

Notes and References

1 Thomson, Thomas, *History of Chemistry*, vol. 1 (London: Henry Colburn
   and Richard Bentley, 1830), p. 307. For a modern treatment, see Donovan,
   Arthur, *Philosophical Chemistry in the Scottish Enlightenment* (Edinburgh:
   Edinburgh University Press, 1975).
2 Simpson, A.D.C. (ed.), *Joseph Black 1728–1799: A Commemorative Symposium* (Edinburgh: Royal Scottish Museum, 1982); Anderson, Robert G.W.
   and Jean Jones (eds), *The Correspondence of Joseph Black*, 2 vols (Farnham:
   Ashgate, 2012).
3 On his work on fixed air, see Guerlac, Henry, 'Joseph Black and fixed air: a
   bicentenary retrospective with some new or little known material', *Isis* 48
   (1957), pp. 124–51 and 433–56.
4 It is instructive to read Black's doctoral dissertation on this subject: Black,
   Joseph, *Experiments upon Magnesia Alba, Quicklime, and other Alcaline
   Substances*, Alembic Club Reprints no.1 (Edinburgh: Alembic Club, 1893);

originally published in Latin in 1754.

5 .On compositionist chemistry, see Chang, Hasok, 'Compositionism as a dominant way of knowing in modern chemistry', *History of Science* 49 (2011), pp. 247–68 and references therein.

6 McKie, Douglas and Niels H. de V. Heathcote, *The Discovery of Specific and Latent Heats* (London: Arnold, 1935) is still the most convenient account of Black's work on heat. Fox, Robert, *The Caloric Theory of Heat from Lavoisier to Regnault* (Oxford: Clarendon Press, 1971) gives a detailed account of the spread of different versions of the caloric theory.

7 On Kirwan's work, see Kim, Mi Gyung, *Affinity, That Elusive Dream* (Cambridge, MA: MIT Press, 2003), pp. 269ff; Taylor, Georgette, 'Tracing influence in small steps: Richard Kirwan's quantified affinity theory', *Ambix* 55 (2008), pp. 209–31. On Richter and the follow-up to his work, see Freund, Ida, *The Study of Chemical Composition* (Cambridge: Cambridge University Press, 1904), ch.7.

8 On this aspect of Lavoisier's work, see Perrin, Carlton, 'Lavoisier's table of the elements: a reappraisal', *Ambix* 20 (1973), pp. 95–105.

9 The most informative survey of this pedagogical tradition remains Anderson, Robert G.W., *The Playfair Collection and the Teaching of Chemistry at the University of Edinburgh 1713–1858* (Edinburgh: Royal Scottish Museum, 1978).

10 Quoted in Anderson, Robert G.W., 'Chemistry beyond the academy: diversity in Scotland in the early nineteenth century', *Ambix* 57 (2010), pp. 84–103 (on p. 91).

11 Davies, Alwyn and Peter Garrett, *UCL Chemistry Department 1828–1974* (St Albans: Science Reviews, 2013) provides an informative account of the London department, which makes an interesting comparison case to the present volume. On the Royal Institution, see Berman, Morris, *Social Change and Scientific Organization: The Royal Institution 1799–1844* (London: Heinemann, 1978).

12 A good sense of the institutional setup of French chemistry may be gained from Donovan, Arthur, *Antoine Lavoisier: Science, Administration, and Revolution* (Oxford: Blackwell, 1993); for the period immediately following Lavoisier's death in 1794, see Crosland, Maurice, *The Society of Arcueil: A View of French Science at the Time of Napoleon I* (London: Heinemann, 1967).

13 Hufbauer, Karl, *The Formation of the German Chemical Community (1720–1795)* (Berkeley and Los Angeles, CA: University of California Press, 1982).

14 On the history of chemistry at the University of Cambridge, there is a comprehensive account: Archer, Mary and Christopher Haley (eds), *The 1702 Chair of Chemistry at Cambridge* (Cambridge: Cambridge University Press, 2005).

15 On Thomson, see Morrell, J.B., 'The Chemist Breeders: the Research Schools of Liebig and Thomas Thomson', *Ambix* 19 (1972), pp. 1–46 .

16 On Liebig, see Brock, William H., *Justus von Liebig: The Chemical Gatekeeper* (Cambridge: Cambridge University Press, 1997). It would be interesting to learn more about the circumstances that led Thomson to set up a system of laboratory-based teaching in Glasgow that was much more akin to Giessen than Edinburgh.

17 A very instructive glimpse at the British scientific community not preoccupied with publication is given in Jungnickel, Christa and Russell McCormmach, *Cavendish: The Experimental Life*, rev. edn (Lewisburg, PA: Bucknell University Press, 1999).

18 For an unorthodox view on this question, see Gillies, Donald, *How Should Research Be Organised?* (London: College Publications, 2008).

19 Addyman's interests are comparable to those of Marcos Martínon-Torres and his colleagues in the archaeological study of alchemical remains. For an introduction, see Martinón-Torres, Marcos and Thilo Rehren, 'Alchemy, chemistry and metallurgy in Renaissance Europe: a wider context for fire assay remains', *Historical Metallurgy* 39 (2005), pp. 14–31.

20 In addition to Donovan, *Philosophical Chemistry in the Scottish Enlightenment* (Edinburgh: Edinburgh University Press, 1975), see Olson, Richard, *Scottish Philosophy and British Physics, 1750–1880: A Study in the Foundations of the Victorian Scientific Style* (Princeton, NJ: Princeton University Press, 1975).

21 For an extensive discussion of more specific lines of 'Newtonianism' in chemistry and physics, see Schofield, Robert E., *Mechanism and Materialism: British Natural Philosophy in an Age of Reason* (Princeton, NJ: Princeton University Press, 1970).

22 A useful starting point, though much neglected nowadays, would be the long chapters on Black and James Watt (as well as those on Priestley and Cavendish) in Crowther, J.G., *Scientists of the Industrial Revolution* (London: Cresset Press, 1962).

Index